T0350958

Graphs and Networks

Graphs and Networks

S. R. Kingan
Brooklyn College and
The Graduate Center,
City University of New York,
New York, NY, USA.

This edition first published 2022
© 2022 John Wiley and Sons, Inc.

The right of S. R. Kingan to be identified as the author of this work has been asserted in accordance with law.

Registered Office
John Wiley & Sons, Inc., 111 River Street, Hoboken, NJ 07030, USA

Editorial Office
111 River Street, Hoboken, NJ 07030, USA

For details of our global editorial offices, customer services, and more information about Wiley products visit us at www.wiley.com.

Wiley also publishes its books in a variety of electronic formats and by print-on-demand. Some content that appears in standard print versions of this book may not be available in other formats.

Limit of Liability/Disclaimer of Warranty
The contents of this work are intended to further general scientific research, understanding, and discussion only and are not intended and should not be relied upon as recommending or promoting scientific method, diagnosis, or treatment by physicians for any particular patient. In view of ongoing research, equipment modifications, changes in governmental regulations, and the constant flow of information relating to the use of medicines, equipment, and devices, the reader is urged to review and evaluate the information provided in the package insert or instructions for each medicine, equipment, or device for, among other things, any changes in the instructions or indication of usage and for added warnings and precautions. While the publisher and authors have used their best efforts in preparing this work, they make no representations or warranties with respect to the accuracy or completeness of the contents of this work and specifically disclaim all warranties, including without limitation any implied warranties of merchantability or fitness for a particular purpose. No warranty may be created or extended by sales representatives, written sales materials or promotional statements for this work. The fact that an organization, website, or product is referred to in this work as a citation and/or potential source of further information does not mean that the publisher and authors endorse the information or services the organization, website, or product may provide or recommendations it may make. This work is sold with the understanding that the publisher is not engaged in rendering professional services. The advice and strategies contained herein may not be suitable for your situation. You should consult with a specialist where appropriate. Further, readers should be aware that websites listed in this work may have changed or disappeared between when this work was written and when it is read. Neither the publisher nor authors shall be liable for any loss of profit or any other commercial damages, including but not limited to special, incidental, consequential, or other damages.

Library of Congress Cataloging-in-Publication Data

Names: Kingan, Sandra, 1966- author.
Title: Graphs and networks / Sandra Kingan.
Description: First edition. | Hoboken, NJ : Wiley, 2022. | Includes
  bibliographical references and index.
Identifiers: LCCN 2021010555 (print) | LCCN 2021010556 (ebook) | ISBN
  9781118937181 (cloth) | ISBN 9781118937204 (adobe pdf) | ISBN
  9781118937273 (epub)
Subjects: LCSH: Graph theory. | System analysis.
Classification: LCC QA166 .K545 2021 (print) | LCC QA166 (ebook) | DDC
  511/.5–dc23
LC record available at https://lccn.loc.gov/2021010555
LC ebook record available at https://lccn.loc.gov/2021010556

Cover Design: Wiley
Cover Image: © Wikipedia

Set in 9.5/12.5pt STIXTwoText by Straive, Chennai, India

To my father Sion Reuben and
in loving memory of my mother Shery Reuben

Contents

List of Figures

Preface

Network science is another name for the modern applications of graph theory and should be studied alongside the usual theorems and techniques in a graph theory course. On the other hand, data scientists, biologists, physicists, and social scientists using networks as models might be interested in the theory behind the results they use. This book attempts to bridge the growing division between graph theory and network science by weaving theory, algorithms, and applications together into the narrative. Only the theorems are numbered because it is through them that the subject unfolds. Moreover, a curated selection of theorems is presented to make the text read like a story, except that this is not fiction and the story never ends.

Chapter 1 presents a broad overview of the book and the basic concepts and examples used throughout the book. Chapter 2 presents foundational topics such as trees, distance, and matrices that represent graphs. Chapters 3 and 4 contain applications found in network science books such as centrality measures, small-world networks, and scale-free networks. Their inclusion early in the book indicates their importance and emphasizes how little classical graph theory these applications require. They do, however, require a working knowledge of elementary linear algebra, probability, and statistics. The required concepts are in Appendices A and B. Chapter 5 and Appendices C and D provide an introduction to graph algorithms including graph traversal algorithms, greedy algorithms, and shortest path algorithms. Subsequently, algorithms are treated as an integral part of the story. Chapter 6 is a collection of results on Eulerian circuits, Hamiltonian cycles, coloring, and higher connectivity including Menger's Theorem and the Splitter Theorem. Chapter 7 provides an introduction to planar graphs. Chapter 8 is on flows, stable sets, matchings, and coverings, and ends with Edmonds' Blossom Algorithm, one of the finest examples of a graph algorithm.

The following two quotes by Jacob Lawrence and Albert Einstein, respectively, capture my approach to writing this book: "when the subject is strong, simplicity is the only way to treat it" and "everything should be made as simple as possible, but not simpler." The book is full of results with short and simple proofs designed to build confidence in reading and understanding proofs. On the other hand, applications such as Google's PageRank require a fairly good grasp of linear algebra. Explaining PageRank without mentioning eigenvalues is certainly possible, however, doing so detracts from the ideas behind this application.

This book is written so that students can review the material and come prepared for a discussion in the classroom. The selection of results in this book gradually increases in difficulty and makes self–study possible. The exercises encourage playing with examples and experimentation. Each chapter has a list of topics suitable for independent study projects and research experiences.

The book also serves as a guide for navigating the large amount of free graph theory and network science resources on the web. Most papers can be obtained within seconds via a web search. Many older textbooks are freely available on the Internet. All these excellent resources for further study are built into the narrative. Students are encouraged to begin gathering original sources as a matter of routine. A seemingly random collection of papers may initially have no connection, but gradually patterns will appear and point toward a research project.

My husband Robert Kingan helped me with the algorithms, figures and references. This book would never have been completed without his encouragement and assistance. My sons Rohan, Arun, and Ravi cheered me on. My colleagues from the New York combinatorics group Kira Adaricheva, Deepak Bal, Nadia Benakli, Jonathan Cutler, Ezra Halleck, Joseph Malkevitch, Eric Rowland, Kerry Ojakian, Peter Winkler, and Mingxian Zhong gave valuable feedback. Many thanks to Noemi Halpern and Murray Hochberg for their long standing support and encouragement and the anonymous reviewers for their nice reviews of my original book proposal. My students helped me by finding typos and errors. I was able to refine explanations by teaching the same topics over and over again. Last, but not least, Susanne Filler, the original acquisitions editor for this book, the expert team at Wiley, Inc. Kimberly Hill, Kalli Schultea, and Gayathree Sekar have been patient and easy to work with. I am grateful to all noted here and to many others who helped in bringing this book to fruition.

My goal in writing this book will be accomplished if students find the material interesting; perhaps interesting enough to pursue research in it. I wrote the book so that chapters can be added ad infinitum. Is your favorite topic missing? Let me know and I'll write a chapter on it and post it online. This is a living and growing book. The book that you hold in your hands is the beginning of a never–ending story.

S. R. Kingan
Brooklyn College and The Graduate Center
The City University of New York
New York, NY

1

From Königsberg to Connectomes

Section 1.1 is a broad introduction to graph theory and network science. Sections 1.2 and 1.3 introduce the nuts and bolts of the subject. Section 1.2 describes when two graphs may be considered the same. Section 1.3 describes ways of obtaining new graphs from old ones and ends with a detailed discussion on minors.

1.1 Introduction

Our story begins in eighteenth century Königsberg, now known as Kaliningrad, a Russian city between Poland and Lithuania. The citizens of Königsberg enjoyed taking walks across the seven bridges over the river Pregel. Some had noticed that they could not cross all the bridges exactly once and return home. Swiss mathematician Leonhard Euler heard about this puzzle and wrote a paper on it called "Solution of a problem relating to the geometry of position" (Euler, 1736). His paper had the simple schematic representation of a map shown in Figure 1.1, where the land masses are labeled A, B, C, D. This drawing subsequently became the point and line drawing that is called a graph. The points are called vertices or nodes and the lines are called edges or links.[1]

Euler explained that if there was a way of walking along the edges of the graph, crossing every edge exactly once, and returning to the starting vertex, then one would have to enter and leave a vertex the same number of times. Therefore every

---

[1] An English translation of Euler (1736) appears in Biggs et al. (1976). In Hopkins and Wilson (2004), the authors note that the diagram on the right in Figure 1.1 made its first appearance in 1892 in W.W. Rouse Ball's book *Mathematical Recreations and Problems of Past and Present Times* (Ball, 1892). According to Parks (2012), the first appearance of the word "graph" is in a note by J.J. Sylvester published in *Nature* (Sylvester, 1878a) followed by a long paper on the same topic in the journal he founded, *American Journal of Mathematics* (Sylvester, 1878b).

Graphs and Networks, First Edition. S. R. Kingan.
© 2022 John Wiley & Sons, Inc. Published 2022 by John Wiley & Sons, Inc.

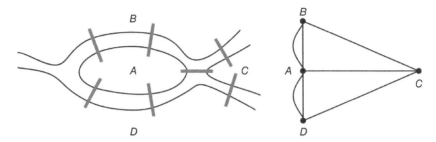

Figure 1.1 The bridges of Königsberg.

vertex must have an even number of edges incident to it. Further, he observed that this necessary condition was also sufficient. Euler did not prove sufficiency. A young German mathematician by the name of Carl Hierholzer gave the first complete proof of this result (Hierholzer, 1873).

Let us change the question slightly and ask "Is there a way of walking along the edges of the graph so as to cross every vertex exactly once, and return to the starting vertex?" This question, posed in the mid-nineteenth century independently by British mathematicians Thomas Kirkman and William Hamilton (Kirkman, 1856; Hamilton, 1858), is still unresolved. There are many theorems giving necessary conditions and many giving sufficient conditions. However, a theorem giving a necessary and sufficient condition, if such a theorem exists, remains elusive.

A *graph* is defined as a finite non-empty set of objects called *vertices* or *nodes* together with a set of unordered pairs of distinct vertices called *edges* or *links*. Typically a graph is denoted by the letter G, the vertex set by $V(G)$ or just V when the context is clear, and the edge set by $E(G)$ or E. The number of vertices is called the *order* of G and the number of edges is called the *size* of G. Throughout the book n stands for the number of vertices and m stands for the number of edges. All this information is summarized by writing $G = (V, E)$.[2]

The graph in Figure 1.2 is called the prism graph since it looks like a flattened rectangular prism viewed from the top. It has vertex set

$$V = \{v_1, v_2, v_3, v_4, v_5, v_6\},$$

---

2 The terminology comes from set theory. The number of elements in a set A is denoted by $|A|$. If x is an element of set A, we write $x \in A$, otherwise we write $x \notin A$. If B is a subset of set A, we write $B \subseteq A$, otherwise we write $B \nsubseteq A$. The empty set ϕ and A itself are called improper subsets of A. All other subsets are called proper subsets. The union of sets A and B, denoted by $A \cup B$, is the set of all elements that are either in A or in B. The intersection of sets A and B, denoted by $A \cap B$, is the set of all elements that are in both A and B. The set of elements in A and not in B is denoted by $A - B$. We assume a set A is a subset of a universal set U. The complement of A, denoted by $\overline{A}$, is the set $(U - A)$. The symmetric difference between two sets A and B is $A \triangle B = (A - B) \cup (B - A) = A \cup B - A \cap B$.

Figure 1.2 The prism graph.

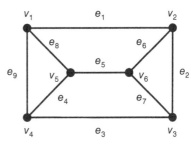

and edge set

$$\{e_1, e_2, e_3, e_4, e_5, e_6, e_7, e_8, e_9\},$$

where $e_1 = (v_1, v_2)$, $e_2 = (v_2, v_3)$, $e_3 = (v_3, v_4)$, $e_4 = (v_4, v_5)$, $e_5 = (v_5, v_6)$, $e_6 = (v_2, v_6)$, $e_7 = (v_3, v_6)$, $e_8 = (v_1, v_5)$, and $e_9 = (v_1, v_4)$. The edges are unordered pairs of vertices, so we may write, for example, $e_1 = v_1 v_2$. Two vertices linked by an edge are called *adjacent vertices*. Two edges with a common vertex are called *adjacent* edges. For example, v_1 and v_2 are adjacent vertices and e_1 and e_2 are adjacent edges. The end vertices of an edge are said to be *incident* to the edge. For example, v_1 is incident to edge e_1.

The *degree* of a vertex v, denoted by $deg(v)$, is the number of edges incident to it. In a graph with n vertices, the degree of each vertex is at most $n - 1$ since each vertex can be joined to at most $n - 1$ of the remaining vertices. The *degree sequence* of a graph is an unordered list of the degrees. The *minimum degree* is denoted by $\delta(G)$ and the *maximum degree* is denoted by $\Delta(G)$. If all the vertices have the same degree, then the graph is called a *regular graph*. For example, the prism graph in Figure 1.2 is a regular graph since every vertex has degree 3. A regular graph of degree 3 is called a *cubic graph*. The *neighborhood* of vertex v, denoted by $N(v)$, is the set of all vertices joined by an edge to v. By convention v is not in $N(v)$.

The graph with just one vertex is called the *trivial graph* and a graph with vertices, but no edges, is called an *isolated graph*. A vertex with no edges incident to it is called an *isolated vertex*.

A *directed graph* (also called *digraph*) is a graph whose edges have a direction. The directed edges, called *arcs*, are ordered pairs of vertices. The arrow on arc (u, v) points from u to v. Vertex u is called the *initial vertex* and vertex v is called the *terminal vertex*. The *outdegree* of vertex v, denoted by $outdeg(v)$, is the number of arcs that have v as the initial vertex. The *indegree* of vertex v, denoted by $indeg(v)$, is the number of arcs that have v as the terminal vertex. The *degree sequence of a digraph* with n vertices, where each vertex v_i has indegree r_i and outdegree s_i, is given by the list of ordered pairs $(r_1, s_1), (r_2, s_2), \ldots, (r_n, s_n)$.

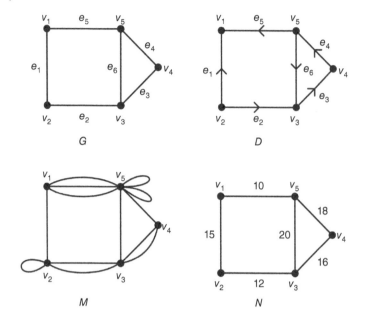

Figure 1.3 An example of a graph, multigraph, digraph, and network.

A *loop* is an edge from a vertex to itself. Multiple edges between a pair of vertices are called *parallel edges*. A *multigraph* is a graph with loops or parallel edges. The Königsburg bridge graph shown in Figure 1.1 is a multigraph.[3]

A graph that has a real number associated with each edge is called a *network*.[4] The numbers are called *weights* or *costs*. Figure 1.3 gives an example of a graph, digraph, multigraph, and network.

A complex network is a real-world network, so large and complicated that it can only be studied empirically. Vertices and edges of a complex network may not be fully known or may be constantly changing. Complex networks are everywhere. Examples include:

- The neurons in the brain linked by synapses (called a connectome);
- The Internet with wired or wireless links between computers;
- The World Wide Web with hypertext links connecting webpages;

---

3 Sometimes a graph is called a simple graph and a multigraph is called a graph. When reading a research paper, the first thing to determine is what exactly the author calls a graph.

4 The term network is used loosely in network science literature to mean graph, digraph, multigraph, network, or some graph–like structure. The rationale is that network should be used when talking about a real-world graph regardless of the type of graph.

- Networks of people where the link is based on some sort of relationship (called social networks); and
- Networks of people where the link is catching a contagious disease such as Ebola or COVID-19 (called epidemic or pandemic networks).

The study of complex networks has become an established interdisciplinary field called Network Science that is populated by physicists, biologists, computer scientists, and social scientists.

A *walk* in a graph is an ordered sequence of vertices and edges

$$v_1, e_1, v_2, e_2, \ldots, v_{k-1}, e_{k-1}, v_k,$$

where $e_{i-1} = v_{i-1}v_i$ is an edge for $2 \le i \le k$.[5] Vertices and edges may be repeated in a walk. A *trail* is a walk with no edge repeated. A *path* is a walk with no vertex repeated. A *closed walk* is one in which the first and last vertex are the same. Closed trails and paths are defined in a similar manner. A closed trail is called a *circuit*. A closed path is called a *cycle*.

For example, in the prism graph shown in Figure 1.2

$$v_1 e_1 v_2 e_2 v_3 e_3 v_4 e_3 v_3 e_7 v_6$$

is a walk from v_1 to v_6. Notice how vertices and edges are repeated. In a trail vertices may be repeated, but not edges. For example,

$$v_2 e_2 v_3 e_7 v_6 e_6 v_2 e_1 v_1$$

is a trail from v_2 to v_1. In a path vertices are not repeated, and consequently edges are not repeated. For example,

$$v_1 e_1 v_2 e_2 v_3 e_3 v_4$$

is a path from v_1 to v_4 and

$$v_1 e_1 v_2 e_2 v_3 e_3 v_4 e_9 v_1$$

is a cycle. For paths and cycles we may use just vertices or just edges since there are no repetitions and therefore no cause for confusion. The aforementioned path may be written as $v_1 v_2 v_3 v_4$ and the cycle may be written as $v_1 v_2 v_3 v_4 v_1$. Alternatively, we may refer to this particular path and cycle by their edges as $e_1 e_2 e_3$ and $e_1 e_2 e_3 e_9$, respectively.

Two walks, trails, or paths are *equal* if they have exactly the same vertices and edges. The *length* of a walk, trail, path, circuit, or cycle is the number of edges in it (counting repetitions when relevant). The *girth* of a graph is defined as the length of the shortest cycle in the graph. A cycle of length 3 is called a *triangle*.

---

5 It is customary to write the phrase "$i \in \{2, 3, \ldots, k\}$" in short as "$2 \le i \le k$," recognizing that in this context i is a natural number.

A circuit that crosses every edge in the graph exactly once is called an *Eulerian circuit*, and a graph with an Eulerian circuit is called an *Eulerian graph*. A cycle that crosses each vertex in the graph exactly once is called a *Hamiltonian cycle*, and a graph with a Hamiltonian cycle is called a *Hamiltonian graph*. As mentioned earlier, Euler gave a complete characterization of graphs with Eulerian circuits, however, a necessary and sufficient condition for the presence of a Hamiltonian cycle is not known.

Suppose a salesman travels to each of n cities exactly once and returns home. The n cities form the vertices of a graph and the routes between cities are the edges. The weights on the edges are the costs of traveling from one city to the other. Consider the problem of finding the cheapest route. For example, in the network shown in Figure 1.4, we could list all possible routes (that would be 5! = 120 routes) to check which one is the cheapest route. This brute-force approach would work for five cities, but what if the traveling salesman is visiting 100 cities? The number of possible routes is 100!. Is there an efficient algorithm? No one knows. This is called the Traveling Salesman Problem, and it is one of the Clay Institute's million dollar problems.[6]

Walks, trails, paths, and cycles in a digraph are defined in a similar manner with the understanding that the direction of the edges must be taken into account while navigating the digraph. A *directed path* in a digraph is an ordered sequence of distinct vertices $v_1, v_2, \ldots, v_k$, where (v_{i-1}, v_i) is an arc for $2 \le i \le k$. A *directed cycle* is a closed directed path.

A non-trivial graph is *connected* if there is a path between every pair of vertices. Otherwise the graph is *disconnected*. By convention the trivial graph is considered to be connected. The *distance* between a pair of distinct vertices u and v in a connected graph G, denoted by $d(u, v)$, is the length of the shortest path between u and v. The *diameter*, denoted by $diam(G)$, is the maximum distance between any pair of vertices. For example, the prism graph shown in Figure 1.2 has diameter 2.

A *connected digraph* is one whose underlying graph is connected. If, for every pair of vertices u and v, there are directed paths from u to v and from v to u, then the digraph is called *strongly connected*. The digraph D in Figure 1.3 is connected,

Figure 1.4 Traveling salesman network.

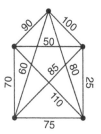

---

6 See https://www.claymath.org/millennium-problems/p-vs-np-problem.

Figure 1.5 Strongly connected digraphs.

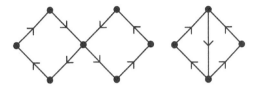

but not strongly connected. There is no directed path from v_1 to v_2. Figure 1.5 displays two strongly connected digraphs.

In a strongly connected digraph the *directed distance* between vertices u and v is the length of the shortest directed path from u to v. Note that in a digraph $d(u, v)$ is not necessarily the same as $d(v, u)$. The diameter of a strongly connected digraph is the maximum distance between any pair of vertices. The first digraph in Figure 1.5 has diameter 4 and the second has diameter 3.

The graph in which every pair of vertices is joined by an edge is called a *complete graph*. The complete graph on n vertices is denoted by K_n (see Figure 1.6). The number of edges in K_n is $\binom{n}{2} = \frac{n(n-1)}{2}$, since every pair of vertices is joined by an edge. Similarly, the digraph in which every pair of vertices u and v is joined by arcs (u, v) and (v, u) is called a *complete digraph*. The number of arcs in a complete digraph on n vertices is $n(n - 1)$.

Next, let us look at some frequently used examples of graphs. These are the toy examples we will use to understand new concepts. The path P_n, for $n \geq 1$, the cycle graph C_n, for $n \geq 3$, and the wheel graph W_{n-1}, for $n \geq 4$ are three families of graphs shown in Figure 1.7. Observe that for P_n the number of edges is $m = n - 1$; for C_n the number of edges is $m = n$; and for W_{n-1} the number of edges is $m = 2(n - 1)$ since the number of spokes in the wheel graph is $n - 1$.

A graph is called *bipartite* if its vertex set V can be written as $V = V_1 \cup V_2$, where V_1 and V_2 are disjoint and every edge joins a vertex in V_1 to a vertex in V_2. A graph is called *complete bipartite* if every vertex in V_1 is joined to every vertex in V_2. A complete bipartite graph with n_1 vertices in V_1 and n_2 vertices in V_2 is denoted by K_{n_1,n_2}.

$K_2 \qquad K_3 \qquad K_4 \qquad K_5 \qquad K_6$

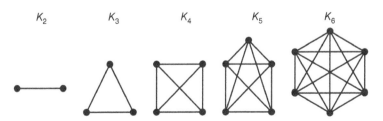

Figure 1.6 Complete graphs.

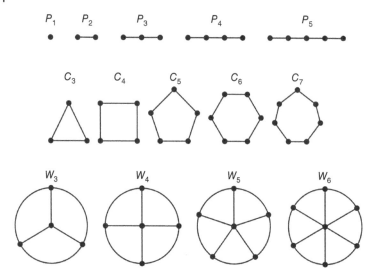

Figure 1.7   Paths, cycles, and wheels.

We can generalize this notion to t-partite graphs, where the vertex set is partitioned into t pairwise disjoint subsets in such a way that every edge joins a vertex in V_i to a vertex in V_j, for $i,j \in \{1, 2, \dots, t\}$ and $i \neq j$. A graph is called *complete t-partite* if every vertex in V_i is joined to every vertex in V_j. A complete t-partite graph with n_i vertices in V_i is denoted by $K_{n_1,\dots,n_t}$. Figure 1.8 shows examples of bipartite and tripartite graphs. The complete bipartite graph $K_{1,n-1}$ with n vertices is called the *star graph*. See Figure 1.9.

Graphs with no cycles are called *acyclic* graphs. A connected acyclic graph is called a *tree*. An acyclic graph is called a *forest*. The terminology is evocative of trees in a forest. The term tree appears in an 1857 paper by British mathematician

Figure 1.8   Bipartite and tripartite graphs.

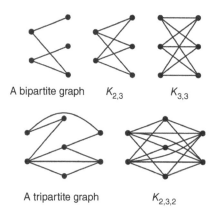

A bipartite graph $K_{2,3}$ $K_{3,3}$

A tripartite graph $K_{2,3,2}$

Figure 1.9 Star graphs.

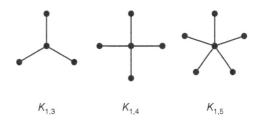

$$K_{1,3} \qquad K_{1,4} \qquad K_{1,5}$$

Arthur Cayley titled "On the theory of the analytical forms called trees" (Cayley, 1857).[7] Cayley was trying to enumerate the isomers of saturated hydrocarbons such as methane, ethane, butane, etc., which have the form C_kH_{2k+2}. An isomer is a molecule with the same chemical formula as another molecule, but with a different arrangement of atoms and different properties. These molecules can be represented by trees as shown in Figure 1.10.

There are many ways to define a substructure of a graph. The most obvious way is to delete some vertices and edges. A graph H is a *subgraph* of a graph G if $V(H) \subseteq V(G)$ and $E(H) \subseteq E(G)$. In the definition of subgraph we understand that if an edge is in $E(H)$, then both its end vertices are in $V(H)$. A *proper subgraph* of G is a subgraph that is not the empty set or G. A subgraph H with $V(H) = V(G)$ is called a *spanning subgraph*. Figure 1.11 displays a graph G and three subgraphs H, I, and J. Observe that J is a spanning subgraph.

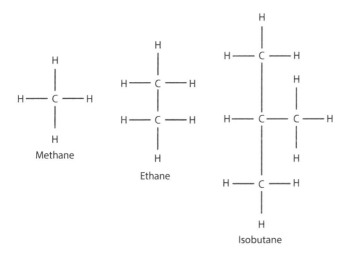

Figure 1.10 Example of trees.

7 According to Biggs et al. (1976) the concept of a tree appears nearly ten years earlier in papers by Kirchhoff (1847) and von Staudt (1847).

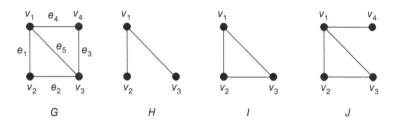

Figure 1.11 Subgraphs.

Sometimes we may want to pick a subset of vertices and look at the subgraph obtained by taking all the edges present that join those vertices. Such a subgraph is said to be "induced" by the subset of vertices. Let $U \subseteq V(G)$. The subgraph induced by U, called an *induced subgraph*, has vertex set U and edge set those edges in G incident with two vertices in U. For example, subgraph I in Figure 1.11 is induced by $\{v_1, v_2, v_3\}$, whereas subgraph H is not an induced subgraph.

A subgraph that is a complete graph is called a *clique*. A *maximal clique* in a graph is a complete subgraph that cannot be made larger by adding more vertices. A *maximum clique* is a maximal clique that has the most number of vertices.

A graph is called *planar* if it can be drawn on the plane without crossing edges. Otherwise it is called *non-planar*. Figure 1.12 gives an example of a planar graph. In the first rendition it is drawn with edges crossing. However, a re-drawing shows that it can be drawn without edges crossing. This graph is K_5 with one edge deleted. What about K_5 and $K_{3,3}$? A few attempts to draw them without crossing edges will reveal that it is impossible. This is because K_5 and $K_{3,3}$ are non-planar.

A map corresponds to a type of planar graph. Vertices can be placed on the countries and two vertices are joined by an edge if the corresponding countries share a common border. Mapmakers have known since antiquity that at most four colors are needed to color two adjacent countries with different colors (countries that meet in a point may be colored the same color). This belief came to be known as the "Four Color Conjecture." It was brought to the attention of British mathematician Augustus de Morgan in 1852 by one of his students Federick Guthrie, who attributed it to his brother Francis Guthrie (Biggs et al., 1976). De Morgan

Figure 1.12 Two drawings of a planar graph.

wrote about it to Hamilton who replied saying he did not find it interesting. In 1976 two American computer scientists from the University of Illinois at Urbana Champaign, Kenneth Appel and Wolfgang Haken, proved the Four Color Conjecture. Their proof is notable in that their mathematical arguments reduced the proof to case-checking of nearly 10 000 cases for which they used a computer. Initially some mathematicians were suspicious of a proof with a computer component, but in time that attitude changed. To commemorate the occasion the university postmark had the slogan "Four colors suffice." A subsequent proof with fewer cases has also been published (Robertson et al., 1997a).

In 1930 Polish mathematician Kazimierz Kuratowski proved that every non-planar graph contains either K_5 or $K_{3,3}$ as a substructure called a subdivision (Kuratowski, 1930). A subdivision of a graph is obtained by placing vertices on edges. Klaus Wagner built on this by proving that every non-planar graph contains either K_5 or $K_{3,3}$ as a substructure called a minor (Wagner, 1937). A minor of a graph is obtained by deleting edges (and any isolated vertices formed as a result) and "contracting edges." An edge is contracted by shrinking it and fusing its end vertices into one vertex. Subdivisions and minors are described in detail in Section 1.3.

A class of graphs $\mathcal{G}$ is said to be *closed under minors* if every minor of a graph in $\mathcal{G}$ is also in $\mathcal{G}$. A graph that is not in $\mathcal{G}$ is called a *minimal excluded minor*[8] if all of its proper minors are in $\mathcal{G}$. Wagner conjectured that every infinite class of graphs closed under minors has a finite set of minimal excluded minors. This conjecture was confirmed by Neil Robertson and Paul Seymour in a series of 23 papers the first of which appeared in 1983 (Robertson and Seymour, 1983) and the last of which appeared in 2010 (Robertson and Seymour, 2010). These papers totaling over 500 pages proved the conjecture as well as dozens of related results and mapped out a thriving research genre.

In the 1930s considerable progress was made in graph theory. In addition to Kuratowski and Wagner's structural result, Karl Menger proved a major connectivity result (Menger, 1927), Frank Ramsey introduced what we now call Ramsey Theory (Ramsey, 1930), Phillip Hall solved the Marriage Problem (Hall, 1935), Hassler Whitney introduced matroids (Whitney, 1935), and Dénes König wrote the first graph theory textbook (König, 1936).[9] Less well known is a graph theoretical development in social psychology. At the 1933 meeting of the New York Medical Society, psychiatrist Jacob L. Moreno presented a graph showing the friendship of boys and girls in a fourth grade class. His work appeared in the New York Times

---

8 In general, a set X is called *minimal* with respect to a certain property if none of its proper subsets have the property. Note that minimal does not mean minimum.
9 In 2021, Martin Charles Golumbic published an annotated translation of Les reseaux (ou graphes), written by Andre Sainte-Lague in 1926. Golumbic called it "The Zeroth Book of Graph Theory."

in an article dated April 3, 1933 titled "Emotions mapped by Geography." He is credited with making a far-sighted statement:

> If we ever get to the point of charting a whole city or a whole nation, we would have a picture of a vast solar system of intangible structures, powerfully influencing conduct, as gravitation does in space. Such an invisible structure underlies society and has its influence in determining the conduct of society as a whole.

Moreno's friendship graph is cited by sociologists as the first example of a social network. His 1934 book *Who Shall Survive: A New Approach to the Problem of Human Interrelations* (Moreno, 1934) has, among other things, several examples of social networks (he called them sociograms) with detailed instructions for drawing them using color, size, and shape of vertices to convey information.

For small graphs it makes sense to list the degrees of the vertices and talk about the degree sequence. For large graphs the list of degrees is too long. So we talk about the frequency distribution of the degrees. The *degree distribution* of a graph is a frequency histogram with the degrees on the x-axis and the number of vertices of each degree on the y-axis. The *indegree distribution* and *outdegree distribution* of a digraph are defined in a similar manner.

In a 1965 study of the citation digraph, where vertices are scientific papers and an arc from u to v indicates u cites v, physicist and historian Derek de Solla Price observed that the degree distribution of the citation digraph followed the power law function, $f(x) = cx^{-n}$, where c and n are constants. Most papers had three to seven citations and only a few papers had many citations (Price, 1965). Graphs whose degree distributions follow the power law function are called *scale-free networks*. The name scale-free comes from the observation that since most vertices have small degree and a few have large degree, the average of the degrees is meaningless.

Meanwhile social psychologist Stanley Milgram conducted his "small-world" experiment, where he randomly selected a group of individuals and gave them letters addressed to someone they didn't know. The goal of the experiment was for each person to send the letter to a person they knew, who in turn sent the letter to a person they knew, and so on. On average Milgram found that the letters reached their destination through six connections (Milgram, 1963). Subsequently these ideas led to the notion of small-world graphs (Watts and Strogatz, 1998).

A centrality measure is a way of quantifying the most important people in a social network. This approach to determining the importance of an individual based on relations with others in the network, and not just on the individual's attributes, was evidently a paradigm shift in sociological thinking. In 1998 Sergey Brin and Larry Page, two Stanford University students, developed a new type of centrality measure called PageRank to determine the importance of a webpage (Brin and Page, 1998). Their company, Google, needs no introduction. In 2004 Harvard student Marc Zuckerberg and his company Facebook made social networks a household term. In 2016 fake news spread through networks of like-minded individuals creating echo-chambers and 2020 brought us the coronavirus pandemic and the unprecedented quarantine and lock-down. Contact-tracing to form the network of people exposed to COVID-19 entered our lexicon. The killing of unarmed black men and women created an uprising and the "Black Lives Matter" movement was born. Predictive-policing based on a smattering of mathematics including graph theory and a large dose of bias appears to have fallen out of favor, for the time being at least.

With this brief initiation into tribal knowledge, let us begin with what is universally considered the first proposition in graph theory.

Proposition 1.1.1. *The sum of the degrees of the vertices in a graph is twice the number of edges.*

Proof. Each edge has exactly two end vertices and therefore contributes exactly two to the sum of the degrees. □

Let G be a graph with n vertices and m edges. We may write Proposition 1.1.1 as

$$\sum deg(v) = 2m.$$

Proposition 1.1.2. *The number of vertices of odd degree is even.*

Proof. Let E be the set of vertices of even degree and O be the set of vertices of odd degree. By Proposition 1.1.1

$$\sum_{v \in E} deg(v) + \sum_{v \in O} deg(v) = 2m.$$

Observe that $\sum_{v \in E} deg(v)$ is even because it is a sum of even numbers. Moreover, $2m$ is clearly even, so $\sum_{v \in O} deg(v)$ must be even. Since each vertex has odd degree, the only way to get an even sum is if the number of odd degree vertices is even. □

Proposition 1.1.2 is called the Handshake Lemma. At a party, consider the people as vertices and assume that two people shaking hands are linked by an edge. We can be certain that the number of people who greet an odd number of acquaintances is even.

Proposition 1.1.3. *In a digraph the sum of the indegrees is equal to the sum of the outdegrees and this sum is equal to the number of arcs.*

Proof. Every arc leaves exactly one vertex and enters exactly one vertex. So every arc contributes one to the sum of the indegrees and one to the sum of the outdegrees. □

Let G be a digraph with n vertices and m arcs. We may write Proposition 1.1.3 as

$$\sum indeg(v) = \sum outdeg(v) = m.$$

The last proposition in this section is a straightforward example of a constructive proof (König, 1936).

Proposition 1.1.4. *Let G be a graph with maximum degree $\triangle(G)$. There exists a regular graph of degree $\triangle(G)$ that contains G as an induced subgraph.*

Proof. If G is a regular graph, then there is nothing to prove. Therefore suppose G is not a regular graph. Let G' be a copy of G and link the corresponding vertices in G and G' whose degrees are less than $\triangle(G)$. Call the resulting graph G_1. If G_1 is regular, then G_1 is the required regular graph containing G as an induced subgraph. If not, let G_1' be a copy of G_1 and link corresponding vertices whose degrees are less than $\triangle(G)$. Continue like this until we arrive at a regular graph G_k of degree $\triangle(G)$, where $k = \triangle(G) - \delta(G)$, as shown in the following diagram.

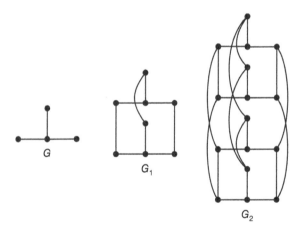

A graph G and a regular graph with G as an induced subgraph.

Note that the proof of Proposition 1.1.4 gives a method for constructing a regular graph containing G as an induced subgraph, but not necessarily the smallest regular graph containing G.

Returning to complex networks, the graph that consists of neurons as vertices and the synapses that connect neurons as edges is called a *connectome*. Depending on the context, it refers to the neurons in the brain or to the neurons in the entire nervous system. The idea that the nervous system is composed of a finite number of nerve cells linked to other cells forming a digraph goes back to Spanish pathologist Santiago Ramón y Cajal in the beginning of the twentieth century. At that time, the "Neuron doctrine" was controversial. Camillo Golgi, inventor of a silver staining technique in 1873, had proposed that the nervous system was a continuous single thread. Cajal proposed a discrete model for the brain of a chicken (Garcia-Lopez et al., 2010). The controversy was settled in favor of Cajal after the electron microscope was invented in 1931.

The first creature whose nervous system was completely mapped was a tiny nematode called *Caenorhabditis elegans* (*C. elegans*). It has 302 neurons that belong to two distinct and independent nervous systems: a large somatic nervous system with 282 neurons (including three isolated neurons that we will disregard) and a small pharyngeal nervous system with 20 neurons. A digraph of the large somatic nervous system with 279 vertices and 2993 directed edges is shown in Figure 1.13 and its indegree and outdegree distributions are shown in Figure 1.14.[10]

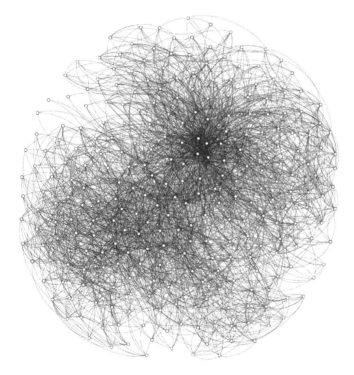

Figure 1.13 *C. elegans* connectome.

---

10 The data is taken from the Worm Atlas Project (https://wormatlas.org/neuronalwiring.html) and drawn in Gephi 0.9.2 using the Fruchterman-Reingold layout. The data includes a

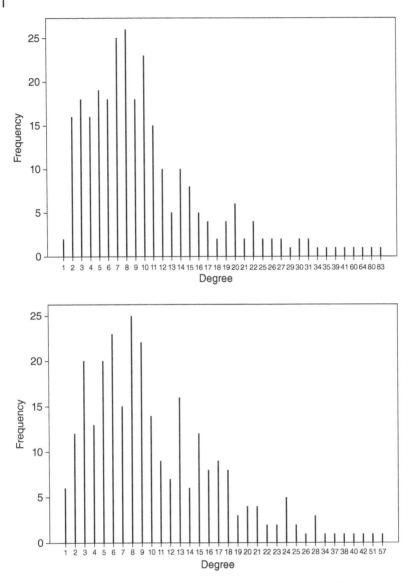

Figure 1.14 *C. elegans* in-degree (top) and out-degree (bottom) distributions.

spreadsheet named NeuronConnect.xls, which lists 6418 links between 279 distinct neurons. Once duplicate edges are removed, 2993 distinct directed links remain. Of the 302 neurons in the *C. elegans* nervous system, 20 are part of the pharyngeal system and are not included, two have no connections, and one connects only to muscle tissue, leaving 279. The cited data source

An examination of a large graph begins with an analysis of its degree distribution since a visual representation may not be of much use. The *C. elegans* connectome is a scale-free network. A few vertices have high degree and most have small degree. Vertices of high degree are called *hubs*. The phenomenon that results in scale-free networks is called *preferential attachment*. The term conveys how newly added vertices connect preferentially to existing vertices maintaining and enhancing hubs.

Let *G* be a graph with *n* vertices and *m* edges. The *mean degree* of *G* (often called *average degree*), denoted by $d(G)$, is given by

$$d(G) = \frac{1}{n} \sum deg(v).$$

It follows from Proposition 1.1.1 that

$$d(G) = \frac{2m}{n}.$$

The *density* of *G* is the number of edges divided by the maximum possible number of edges in the graph. In symbols,

$$density(G) = \frac{m}{\binom{n}{2}} = \frac{2m}{n(n-1)}.$$

The corresponding definitions for a digraph are similar except that the maximum number of arcs in a digraph is $n(n-1)$. Let *D* be a digraph with *n* vertices and *m* arcs. The mean degree of *D* is

$$d(D) = \frac{m}{n}$$

and the density of *D* is

$$density(D) = \frac{m}{n(n-1)}.$$

For example, the *C. elegans* connectome (the large somatic nervous system with 279 vertices and 2993 directed edges) has mean degree and density as follows:

$$d(G) = \frac{m}{n} = \frac{2993}{279} \approx 10.728,$$

$$density(G) = \frac{m}{n(n-1)} = \frac{2993}{279(279-1)} \approx 0.039.$$

Other connectomes can be analyzed in a similar manner.

In an epidemic network or a pandemic network, the nodes are living entities like people, animals, or plants and a directed edge from one node to another indicates transmission of a contagion. The average degree is denoted by R_0 and is called the *reproduction number* because when a disease is transmitted to a person

---

does not include data on the pharyngeal system; that data can be obtained from the WormWeb project, at http://wormweb.org/details.html.

it reproduces itself. In order for an outbreak to show signs of ending, R_0 must become less than 1. For example, the 2009 outbreak of the H1N1 virus (swine flu) had an R_0 value between 1.4 and 1.6, whereas the 1918 outbreak had an R_0 value between 1.4 and 2.8. Seasonal strains of influenza have an R_0 value between 0.9 and 2.1 (Coburn et al., 2009). The jury is still out on the R_0 value of the COVID-19 pandemic network.[11]

We will end this section with some additional information on connectomes. The function of a neuron is to receive information from some neurons, process that information, and send it to other neurons. A neuron consists of three parts: the cell body (soma); the axon, which looks like a long tail and transmits messages from the nucleus; and the dendrites which look like fine branches of a tree and receive messages for the nucleus.[12] Cajal classified neurons based on the shapes and sizes of the nucleus, axon, and dendrites. He compared the shapes of neurons that he saw to forms in nature such as flowers, trees, branches, roots, etc. and was the first to recognize that neurons were discrete members of what he called the "garden of neurons." However, Sebastian Seung writes in his book *Connectome: How the brain's wiring makes us who we are*, that Cajal's approach to classifying neurons was "much as a nineteenth-century naturalist would have classified different species of butterflies" (Seung, 2012). Seung raises the possibility that there may be another way of classifying neurons.

The brain is divided into different regions based on function. The process of identifying parts of the brain involved in different functions dates back to 1891 when French neurosurgeon Paul Broca and German neurologist Carl Wernicke located the region in the brain involved in language processing. Broca's area, located in the lower part of the frontal lobe, is involved in the formation of sentences. Wernicke's area, located in the temporal lobe, is involved in the comprehension of speech. Wernicke hypothesized the existence of a bundle of long axons connecting Broca's and Wernicke's regions. He predicted that damage to these links will leave both speech production and comprehension intact, yet it would be impossible for a person to hear someone talking and repeat their words. Subsequently, several patients with this problem called "conduction aphasia" were found. Thus Wernicke established that links between regions are just as important as the regions.[13]

The network consisting of regions as vertices and links between regions as edges is called the *regional connectome*. The goal of the Human Connectome Project

---

11 The article "A guide to R - the pandemic's misunderstood metric" describes the pitfalls of using R_0 in making policy (Adam, 2020). See also "Case clustering emerges as key pandemic puzzle" (Kupferschmidt, 2020).
12 Robert Stufflebaum's *Mind project* located at http://www.mind.ilstu.edu/.
13 See http://thebrain.mcgill.ca.

launched in 2010 by the National Institutes of Health is to obtain a thorough understanding of the regional connectome. It is however a shortcut, Seung says, describing the approach of dividing the brain into regions as follows:

> Neuroscientists continue to argue over their classification. Their disagreements are a sign of a more fundamental problem: It's not even clear how to properly define the concepts of "brain region" and "neuron type." In Plato's dialogue *Phaedrus*, Socrates recommends "division…according to the natural formation, where the joint is, not breaking any part as a bad carver might." This metaphor vividly compares the intellectual challenge of taxonomy to the more visceral activity of cutting poultry into pieces. Anatomists follow Socrates literally, dividing the body by naming its bones, muscles, organs, and so on. Does his advice also make sense for the brain?

Seung proposes developing a regional classification of the brain based on the manner in which neurons are linked. He says linked groups of neurons have a common "connectional fingerprint," and these substructures could then be used to define the "regions" of the brain. It is an intriguing idea; one that may benefit from the application of graph structure theorems.

1.2 Isomorphism

We say two graphs G_1 and G_2 are *isomorphic* if there is a bijective function (a one-to-one and onto function) from $V(G_1)$ to $V(G_2)$ that preserves adjacencies. Otherwise they are *non-isomorphic*. In symbols we write $G_1 \cong G_2$, if there is a bijective function f from $V(G_1)$ to $V(G_2)$ such that

$$(u, v) \in E(G_1) \text{ if and only if } (f(u), f(v)) \in E(G_2).$$

Figure 1.15 shows three pairs of isomorphic graphs: $G_1 \cong G_2$, $G_3 \cong G_4$, and $G_5 \cong G_6$. It is easy to see G_1 is isomorphic to G_2. The bijective function f from $V(G_1)$ to $V(G_2)$ given by $f(1) = b, f(2) = c, f(3) = d, f(4) = a$ is an isomorphism because it preserves adjacencies. Preserving adjacencies means edge $(1, 2)$ is mapped to edge (b, c), edge $(2, 3)$ is mapped to edge (c, d), edge $(3, 4)$ is mapped to edge (d, a), edge $(1, 4)$ is mapped to edge (b, a), and edge $(1, 3)$ is mapped to (b, d).

The isomorphism between the second pair G_3 and G_4 is less obvious. The bijective function g from $V(G_1)$ to $V(G_2)$ given by $g(1) = a, g(2) = c, g(3) = e, g(4) = b, g(5) = d$, and $g(6) = f$ is the required isomorphism.

The isomorphism between the third pair G_5 and G_6 is given by the bijective function h from $V(G_1)$ to $V(G_2)$ given by $h(1) = a, h(2) = e, h(3) = c, h(4) = b, h(5) = d$.

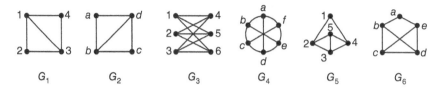

Figure 1.15 Three pairs of isomorphic graphs.

The relation "is isomorphic to" is an equivalence relation on the set of graphs.[14] To see this, observe that isomorphism is:

(1) Reflexive since a graph is isomorphic to itself by the identity function.
(2) Symmetric since if $G_1 \cong G_2$ by the bijection f, then $G_2 \cong G_1$ by the inverse function f^{-1} that exists and is a bijection.
(3) Transitive since if $G_1 \cong G_2$ and $G_2 \cong G_3$ by the bijections f and g, respectively, then $G_1 \cong G_3$ by the composite function $g \circ f$ that exists and is a bijection.

Isomorphism partitions the set of graphs into subsets called equivalence classes.[15] Two graphs G_1 and G_2 are in the same equivalence class if and only if $G_1 \cong G_2$. All graphs in the same equivalence class are isomorphic to each other and any one graph from the class serves to represent the entire class. Figure 1.16 shows the non-isomorphic graphs on 1, 2, 3, and 4 vertices.

An isomorphism from a graph to itself is called an *automorphism*. For example, consider the graph G_1 in Figure 1.15. The function $f : \{1, 2, 3, 4\} \longrightarrow \{1, 2, 3, 4\}$ given by $f(1) = 3, f(2) = 4, f(3) = 1, f(4) = 2$ is an automorphism of G_1.

An *invariant* of a graph G is a property that is the same for any graph isomorphic to G. For example, the number of vertices, the number of edges, and the degree sequence are graph invariants. To determine if two graphs G_1 and G_2 are isomorphic, we first check as many invariants as possible to see if they differ on G_1 or G_2. However, if all the known invariants match, there is no guarantee that the two graphs are isomorphic until we find the bijection that preserves adjacencies.

---

14 The Cartesian product of two sets A and B, denoted by $A \times B$, is the set of ordered pairs (a, b), such that $a \in A$ and $b \in B$. We write $A \times B = \{(a, b) \mid a \in A, \ b \in B\}$. A relation R is a subset of $A \times B$. A relation R is reflexive if $(x, x) \in R$; symmetric if $(x, y) \in R \iff (y, x) \in R$; antisymmetric if $(x, y) \in R$ and $(y, x) \in R \implies x = y$; and transitive if $(x, y) \in R$ and $(y, z) \in R \implies (x, z) \in R$. An equivalence relation on a set S is a relation $\sim$ on $S \times S$ that is reflexive, symmetric, and transitive. For example, the equality relation on the set of integers $\mathbb{Z}$ is an equivalence relation.
15 A partition of a set S is a family of pairwise disjoint subsets of S whose union is S. An equivalence relation imposes a partition on S, where each subset consists of elements that are related to each other. Conversely, every partition of S gives rise to an equivalence relation $\sim$, where $x \sim y \iff x$ and y lie in the same sub set of the partition.

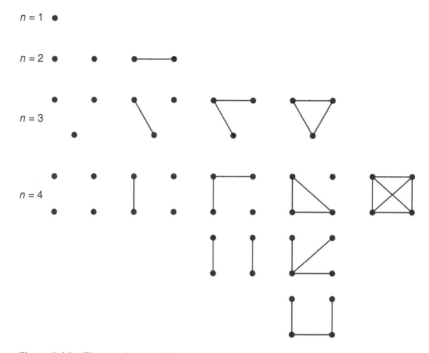

Figure 1.16 The non-isomorphic graphs on $n \leq 4$ vertices.

The "brute-force" way to do this would be to check $n!$ possible mappings in the worst case.

Next, let us study another definition of "sameness." A graph on n vertices with distinct labels attached to the vertices such as $v_1, \ldots, v_n$ or simply $1, 2, \ldots, n$ is called a *labeled graph*. Two labeled graphs G_1 and G_2 are *label isomorphic* or *identical* if there is a bijection from $V(G_1)$ to $V(G_2)$ that preserves labeled adjacencies. Otherwise they are *non-identical*. Figure 1.17 shows the non-identical graphs on 1, 2, and 3 vertices.

The next result determines precisely the number of non-identical graphs on n vertices.

Proposition 1.2.1. *The number of non-identical graphs on n vertices is $2^{\binom{n}{2}}$.*

Proof. The maximum number of edges in a graph with n vertices is $\binom{n}{2}$. Each edge may or may not be in the graph. So there are $2^{\binom{n}{2}}$ non-identical graphs. [16] $\square$

---

16 There are some combinatorial ideas embedded in the argument worth elucidating, especially the well-known combinatorial identity $\sum_{k=0}^{t} \binom{t}{k} = 2^t$. To prove this identity, count the same thing

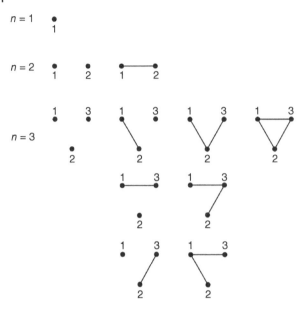

Figure 1.17 The non-identical graphs on $n \leq 3$ vertices.

Isomorphism is not quite relevant for large graphs in the sense that it is unlikely that two large graphs arising from real world situations would be isomorphic. One approach adopted by network scientists is to compare and contrast complex networks based on their rough shapes. Finding the approximate shape of a complex network is no small task. For example, the World Wide Web (the digraph formed with webpages as vertices, where two pages are joined by an arc if one page links to another) evidently has a distinct bow-tie shape according to Broder et al. (2000).[17] Bacterial metabolic networks (digraphs consisting of chemical reactions and their links to each other) also exhibit a bow-tie shape (Csete and Doyle, 2004). Figure 1.18 displays some typical shapes of large graphs.

---

in two different ways. Let S be a set of t elements. The expression on the left counts the number of k-element subsets of S, where k ranges from 0 to t. The number of subsets of S can be counted in a different way. Let X be a subset of S. Each element of S can be inside X or outside X, leading to two choices per element. So the number of ways of constructing X is 2^t. In the case of graphs, the number of non-identical graphs with n vertices and m edges, where $m \leq n$ is $\binom{\binom{n}{2}}{m}$. To obtain the total number of non-identical graphs on n vertices sum over all possible m and use the combinatorial identity to get $\sum_{m=0}^{\binom{n}{2}} \binom{\binom{n}{2}}{m} = 2^{\binom{n}{2}}$.

17 Later research showed that the bow-tie shape of the Web graph was not entirely correct. While the Web graph does have a large central core, the bow-tie structure was the result of the Web crawling techniques used (Meusel et al., 2015).

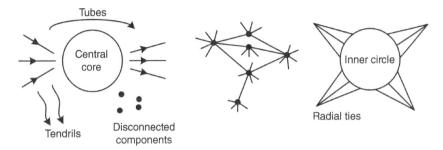

Figure 1.18 Shapes of graphs.

Easley and Kleinberg (2010) describe how the spread of email recommendations for a Japanese comic book and the spread of a tuberculosis outbreak have a similar appearance (see the second diagram in Figure 1.18). Social contagions, unlike biological contagions, involve decision-making by the individual. However, as Easley and Kleinberg note "the network-level dynamics are similar, and insights from the study of biological epidemics are also useful in thinking about the processes by which ideas spread on social networks."

In a *collaboration graph* people are represented as vertices and two people are joined by an edge if they collaborated on a project. In mathematics a famous collaboration graph revolves around Paul Erdős, who wrote hundreds of papers with collaborators. The 511 mathematicians who coauthored a paper with Erdős are assigned Erdős number 1, mathematicians who coauthored a paper with a coauthor of Erdős are assigned Erdős number 2, and so on. Erdős himself is assigned Erdős number 0. An edge is drawn between two mathematicians if they collaborated on a paper together (possibly they jointly coauthored a paper with Erdős). The Erdős collaboration graph is evolving, although the number of mathematicians with Erdős number 1 is fixed at 511 since Erdős died in 1996.

The shape of this collaboration graph looks like the third diagram in Figure 1.18. It graph has a densely connected core along with loosely coupled radial branches reaching out from the core. The close knit inner circle of collaborators share a common "mathematical tongue" that is critical to proving new results. The radial ties bring diversity in the form of new collaborators.[18]

The entire Erdős-1 collaboration graph is presented in Figure 1.19. It is drawn using Gephi 0.9.2 using the Force-Atlas layout algorithm. Vertices with higher degree are darker. Note that Erdős does not appear in the Erdős-1 collaboration

---

18 Jerold Grossman and his collaborators created a website on the Erdős Number Project located at http://www.oakland.edu/enp/. The Erdős -1 collaboration graph and its description were originally prepared by Valdis Kreb based on Grossman's data and is located at http://www.orgnet.com/Erdos.html.

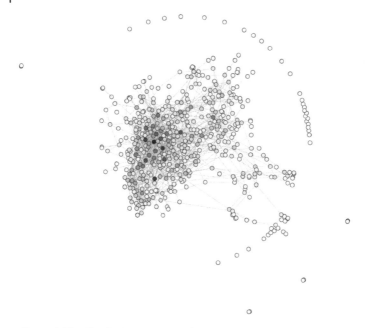

Figure 1.19   The Erdős-1 collaboration graph.

graph and coauthors of Erdős who did not publish a paper with any other coauthor of Erdős appear as isolated vertices in the graph.

We will end this section with the Reconstruction Conjectures. The subgraph obtained by deleting one vertex v is called a *vertex-deletion* and is denoted by $G - v$. The subgraph obtained by deleting one edge e is called an *edge-deletion* and is denoted by $G\backslash e$. When a vertex is deleted, all the edges incident with the vertex are removed. When an edge is deleted, only the link between the two end vertices is removed, while the end vertices remain. Any isolated vertices that may be formed remain intact. If this is not the case it will be specified.

Suppose G is a graph with n vertices labeled $v_1, \ldots, v_n$. Consider the labeled subgraphs of vertex-deletions: $G - v_1, G - v_2, \ldots, G - v_n$. This multiset of vertex-deletions is called the *deck* of vertex-deletions. If you are shown only the deck of vertex-deletions, can you reconstruct G? The Vertex Reconstruction Conjecture states that two graphs (with at least 3 vertices) that have the same deck of vertex-deletions are isomorphic (Kelly, 1957; Ulam, 1960). Figure 1.20 displays two non-isomorphic graphs on 3 vertices and their decks of vertex-deletions. Observe that the decks are different.

Now suppose we remove all vertex labels from the vertex-deletions and keep only non-isomorphic graphs. This set of unlabeled subgraphs is called the set of vertex-deletions. The vertex-deletions in Figure 1.20, although different when

Figure 1.20 Two non-isomorphic graphs and their decks of vertex-deletions.

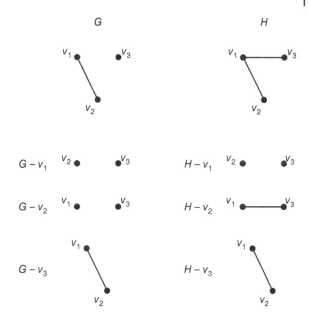

viewed as multisets, are the same when viewed as sets since $G - v_1 \cong G - v_2$ and $H - v_2 \cong H - v_3$. Can G be reconstructed from its set of vertex-deletions? This harder question, known as the *Set Reconstruction Conjecture*, was proposed by Harary (1964).

One approach to tackling the Vertex Reconstruction Conjecture is to see how much information about the graph can be recovered from its deck of vertex-deletions. Proposition 1.2.2 lists three graph invariants that can be reconstructed.

Proposition 1.2.2. *The number of vertices, edges, and the degree sequence of a graph can be reconstructed from its deck of vertex-deletions.*

Proof. Let G be a graph with n vertices $v_1, \ldots, v_n$ and m edges. The number of vertices is clearly reconstructable since it is the number of graphs in the deck. To see that the number of edges of G is reconstructable, let m_i be the number of edges in $G - v_i$, where $1 \leq i \leq n$. Each edge appears in $n - 2$ of the vertex-deletions. Therefore the number of edges is

$$m = \frac{1}{n - 2} \sum_{i=1}^{n} m_i.$$

To find $deg(v_i)$, observe that in $G - v_i$, the missing edges are precisely $deg(v_i)$. Therefore $deg(v_i) = m - m_i$. □

Another approach to tackling the Vertex Reconstruction Conjecture is to prove that it holds for special classes of graphs; for instance regular graphs. In fact, if G is a regular graph of degree t, it can be reconstructed from a single vertex-deletion $G - v$. To see this, observe that the vertices in $G - v$ are of two types: those with degree t and those with degree $t - 1$. Moreover, there will be t vertices of degree $t - 1$. To obtain G add a new vertex adjacent to the t vertices of degree $t - 1$.

The Edge Reconstruction Conjecture is analogous to the Vertex Reconstruction Conjecture. Suppose G is a graph with m edges labeled $e_1, \dots, e_m$. Consider the labeled subgraphs of edge-deletions: $G\backslash e_1, G\backslash e_2, \dots, G\backslash e_m$. This multiset of edge-deletions is called the deck of edge-deletions. If you are shown only this deck of edge-deletions, can you reconstruct G? The Edge Reconstruction Conjecture states that graphs (with at least 6 edges) that have the same decks of edge-deletions are isomorphic (Harary, 1964). A corresponding conjecture also exists for sets of edge-deletions.

1.3 Constructions and Minors

We begin this section by describing three ways of combining two graphs G_1 and G_2 to obtain a new graph. The *join* of G_1 and G_2, denoted by $G_1 + G_2$, has vertex set $V(G_1) \cup V(G_2)$ and every vertex in $V(G_1)$ is joined to every vertex in $V(G_2)$. To construct the join draw the two graphs next to each other and connect every vertex in one to every vertex in the other. For example, the wheel graph W_{n-1}, shown in Figure 1.7, may be viewed as the join of the single vertex and the cycle C_{n-1} with $n - 1$ vertices. Figure 1.21 gives another example of a join.

The *Cartesian product* of two graphs G_1 and G_2, denoted by $G_1 \times G_2$, has vertex set

$$V(G_1) \times V(G_2) = \{(v_i, v_j) \mid v_i \in V(G_1), \ v_j \in V(G_2)\}.$$

Two vertices (v_i, v_j) and (v_k, v_l) are adjacent if $v_i v_k \in E(G_1)$ and $j = l$ or $v_j v_l \in E(G_2)$ and $i = k$. Figure 1.22 displays the Cartesian product of the path graphs P_m and P_n and the Cartesian product of P_3 and the cycle C_3.

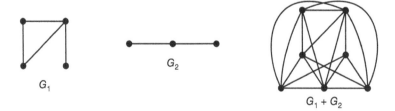

Figure 1.21 An example of a join.

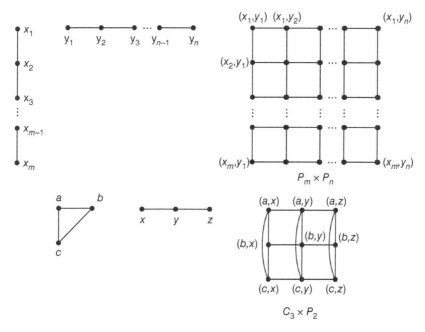

Figure 1.22 Two examples of Cartesian products.

Using the Cartesian product we can define the n-dimensional "cubes," Q_n, recursively as: $Q_1 = P_2$; and for $n \geq 2$, $Q_n = Q_{n-1} \times P_2$. Figure 1.23 gives a nice visualization of the four-dimension analog of the cube, called a tesseract. Start with P_2 whose vertices are labeled a and b. To obtain $Q_2 = Q_1 \times P_2$ draw two copies of P_2

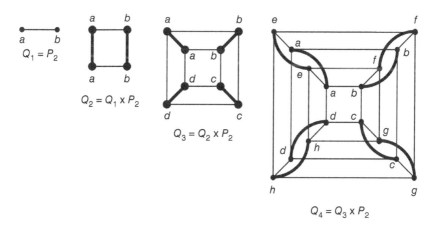

Figure 1.23 The four-dimensional cube.

next to each other and join corresponding vertices. The edges joining corresponding vertices are shown in bold. To obtain $Q_3 = Q_2 \times P_2$ draw two copies of Q_2 and join corresponding vertices. Similarly, to obtain $Q_4 = Q_3 \times P_2$ draw two copies of Q_3 and join corresponding vertices. We can assign coordinates to each of the vertices so that each adjacent vertex differs by exactly one coordinate. The vertices of Q_2 may be labeled $(0, 0)$, $(1, 0)$, $(0, 1)$, and $(1, 1)$. Similarly, the vertices of Q_3 may be assigned 3-tuples, and the vertices of Q_4 may be assigned 4-tuples. Viewed in this manner Q_4 is the four-dimensional cube.

Suppose each of G_1 and G_2 has a clique K_t, where $t \geq 1$. The *clique sum* of G_1 and G_2 across K_t is formed by placing two copies of K_t, one on top of the other, and identifying the vertices and edges. For example, the clique sum of G_1 and G_2 across K_1 (a vertex) is obtained by attaching G_1 and G_2 at the specified vertex. The clique sum across K_2 (an edge) is obtained by attaching G_1 and G_2 along the specified edge. The clique sum across K_3 (a triangle) is obtained by attaching G_1 and G_2 along the specified triangle, and so on. For $t \geq 2$, if in addition the identified edges of K_t are removed, then the clique sum is called a *t-sum*.[19] Figure 1.24 gives examples of clique sums across K_2 and K_3 with all the edges removed.

The *complement* of a graph G, denoted by $\overline{G}$, is the graph that has vertex set $V(G)$ and a pair of vertices is joined by an edge in $\overline{G}$ if there is no edge joining them in G. Figure 1.25 gives three examples of graphs and their complements. Some graphs like P_4 and C_5 are isomorphic to their complements. Such graphs are called *self-complementary*. The next result gives a necessary condition for a graph to be self-complementary.

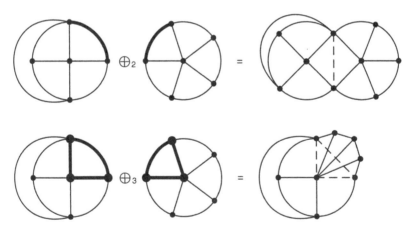

Figure 1.24 $(K_5 \backslash e) \oplus_2 W_5$ and $(K_5 \backslash e) \oplus_3 W_5$.

---

19 The terms "clique sum" and "*t*-sum" depend on how authors define them. Some authors allow a subset of the identified edges to stay.

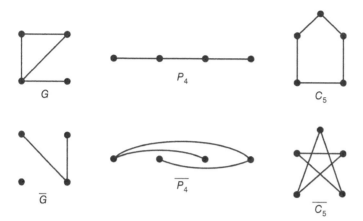

Figure 1.25 Examples of complements.

Proposition 1.3.1. *If G is a self-complementary graph on n vertices, then n = 4k or n = 4k + 1, for some integer k ≥ 1.*

Proof. Suppose G has m edges and let $\overline{G}$ be its complement. Observe that

$$|E(G)| + |E(\overline{G})| = \frac{n(n-1)}{2}.$$

Since G is self-complementary, $|E(G)| = |E(\overline{G})| = m$. So $2m = \frac{n(n-1)}{2}$ and $m = \frac{n(n-1)}{4}$. Since m is a natural number, n or $n-1$ must be a multiple of 4. Therefore $n = 4k$ or $n = 4k + 1$, for some integer $k \geq 1$. □

A *subdivision* of a multigraph G is obtained from G by replacing edges with paths of length at least 2 and loops by cycles. Figure 1.26 shows a multigraph G and a graph H and their subdivisions.

A *series-parallel network* is a multigraph obtained from a loop or a single edge by subdividing edges or adding edges in parallel. Subdivisions and series-parallel networks are related, but not exactly the same. A series-parallel network begins with a loop. Edges are added in parallel to existing edges and vertices are placed on edges. A subdivision may be constructed from any graph H, by placing vertices on edges. For example, the subdivision of the multigraph G in Figure 1.26 is a series-parallel network. However, the subdivision of graph H is not a series-parallel network.

Let $S = \{S_1, S_2, \ldots, S_n\}$ be a family of sets. The *intersection graph* of S is the graph formed by treating the n sets in S as vertices and joining two vertices by

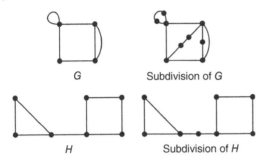

Figure 1.26 Examples of a subdivision.

an edge if their corresponding sets have a non-empty intersection. If the sets are intervals of real numbers, then the intersection graph is called an *interval graph*.

The *line graph* of a graph G, denoted by $L(G)$, is obtained from G by placing a vertex on each edge and joining vertices if the corresponding edges are adjacent. Figure 1.27 gives an example of a graph and its line graph. The presence of isolated vertices in G has no impact on $L(G)$. Moreover, the line graph of a disconnected graph consists of the line graphs of each connected component. Thus we may assume the graph is connected. The next result gives the number of vertices and edges in a line graph.

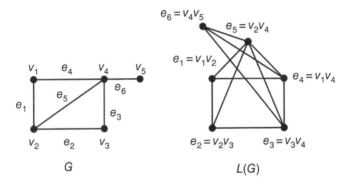

Figure 1.27 Line graphs.

Proposition 1.3.2. *Let G be a connected graph with $n \geq 2$ vertices and $m \geq 1$ edges. The number of vertices in $L(G)$ is m and the number of edges is $-m + \frac{1}{2}\sum_{v \in V}[deg(v)]^2$.*

Proof. By definition the number of vertices in $L(G)$ is m. To obtain the number of edges, observe that each of the $\binom{deg(v)}{2}$ pairs of edges adjacent to v gives rise to an edge in $L(G)$. Therefore the number of edges in $L(G)$ is $\sum_{v \in V}\binom{deg(v)}{2}$. Simplifying this expression gives

$$\sum_{v \in V} \binom{deg(v)}{2} = \frac{1}{2} \sum_{v \in V} deg(v)[deg(v) - 1] = \left[\frac{1}{2} \sum_{v \in V} [deg(v)]^2 \right] - m. \qquad \square$$

A graph G is a *line graph* if $G \cong L(H)$ for some graph H. Line graphs have several interesting properties. For example, each vertex in G is in at most two cliques. To see this, observe that the edges incident to a vertex in H form a complete subgraph in G. Since each edge of H has two end vertices, it is in the neighborhood of exactly two vertices. Therefore each vertex in G is in at most two cliques. Not all graphs are line graphs. For example, the star graph $K_{1,3}$ is not a line graph, since the vertex that is joined to the other three vertices is in three cliques.

A class of graphs $\mathcal{G}$ is said to be *closed under induced subgraphs* if every induced subgraph of a graph in $\mathcal{G}$ is also in $\mathcal{G}$. For example, an induced subgraph of a line graph is also a line graph. A *forbidden induced subgraph* is a graph that is not in $\mathcal{G}$, but all of its proper induced subgraphs are in $\mathcal{G}$. *A priori* there is no reason to believe the list of forbidden induced subgraphs for line graphs is finite. However, Lowell Beineke proved precisely such a result (Beineke, 1970), which we will state without proof.

Theorem 1.3.3. *Let G be a connected graph. Then $G = L(H)$ for some graph H if and only if G has no induced subgraph isomorphic to one of the nine graphs in Figure 1.28.*

Minors are substructures that are a bit more complicated than subgraphs. First we must understand the contraction operation. To *contract* an edge e with end vertices u and v, collapse the edge and identify the end vertices u and v as one vertex. The resulting graph, denoted by G/e, is called an *edge-contraction*. Note that parallel edges and loops may be formed in the contraction process. A graph H is a *minor* of a graph G, if H can be obtained from G by deleting edges or contracting edges. We write $H = G\backslash X/Y$, where X is the set of edges deleted and Y is the set of edges contracted. The focus is entirely on edges. Figure 1.29 shows examples of deletions and contractions.

An edge with an end-vertex of degree 1 is called a pendant edge. If e is a loop or a pendant edge, then $G\backslash e = G/e$. If H is obtained from G only by deleting edges and any resulting isolated vertices, then H is called a *deletion-minor* of G. Similarly, if H is obtained from G only by contracting edges, then H is called a *contraction-minor* of G.

A minor is not necessarily a subgraph. Consider the Petersen graph P shown in Figure 1.30 with 10 vertices and 15 edges.[20] It does not contain K_5 as a subgraph

---

20 The Petersen graph is a useful example to test any conjectures that come to mind. Julius Petersen attributes it to J.J. Sylvester Petersen (1898). It appears in Kempe (1886) for the first time according to Biggs et al. (1976).

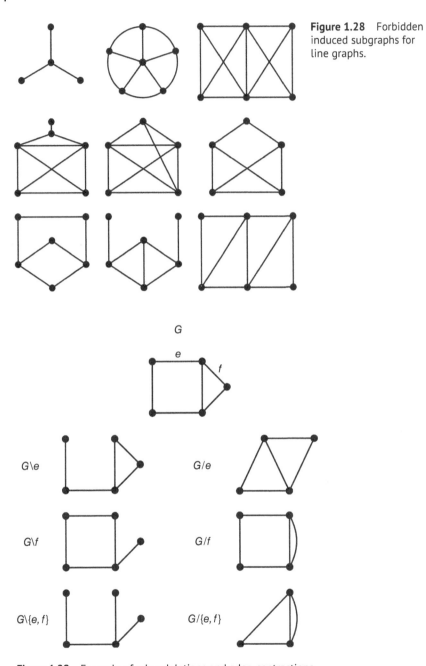

Figure 1.28 Forbidden induced subgraphs for line graphs.

Figure 1.29 Example of edge-deletions and edge-contractions.

Figure 1.30 Petersen graph.

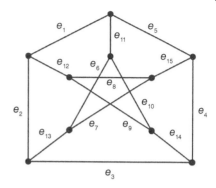

since it has no vertex of degree 4. However, it does contain K_5 as a minor. To see this, contract the five edges e_{11}, e_{12}, e_{13}, e_{14}, and e_{15} that join the outer pentagon to the inner star; the resulting minor is K_5. We write

$$P/\{e_{11}, e_{12}, e_{13}, e_{14}, e_{15}\} \cong K_5.$$

A subgraph is not necessarily a minor. If the subgraph is formed only by deleting edges then the isolated vertices stay, and the subgraph is not a minor. This is the only situation where a subgraph is not a minor.

Let $\mathcal{M}$ be a class of graphs closed under minors. That is, if $G \in \mathcal{M}$, then $G\backslash e \in \mathcal{M}$ and $G/e \in \mathcal{M}$ for every edge e of G. Recall from Section 1.1 that a minimal excluded minor for $\mathcal{M}$ is a graph that is not in $\mathcal{M}$, but all of its proper minors are in $\mathcal{M}$. For example, the class of forests is closed under minors, since a minor of a forest is also a forest. The class of trees, however, is not closed under minors since removal of an edge in a tree results in a disconnected graph. The next result is an example of a straightforward excluded minor result.

Proposition 1.3.4. *A graph G is a forest if and only if G has no minor isomorphic to K_3.*

Proof. The smallest graph that is not a forest is K_3. Since the class of forests is closed under minors, if G is a forest, then G has no K_3-minor. Conversely, suppose G has no K_3-minor. Suppose, if possible, G is not a forest. Then G has a cycle C. A minor isomorphic to K_3 may be obtained by deleting all the edges outside C (and any isolated vertices formed as a result) and contracting all except three edges of C. This is a contradiction to the hypothesis. Therefore G is a forest. $\qquad\square$

The class of series-parallel networks is closed under minors and has an excluded minor characterization: A connected graph is a series parallel network if and only if it has no minor isomorphic to K_4. It is easy to see that K_4 is a minimal excluded minor for series-parallel networks by drawing all the connected series-parallel networks up to 4 vertices. The other direction requires additional results and appears in Section 6.5.

The set of graphs $\mathcal{G}$ with the relation "is a minor of" is a partially ordered set.[21] To see this, observe that the minor relation is:

- Reflexive since a graph is a minor of itself.
- Antisymmetric since if G_1 is a minor of G_2 and G_2 is a minor of G_1, then $G_1 \cong G_2$.
- Transitive since if G_1 is a minor of G_2 and G_2 is a minor of G_3, then G_1 is a minor of G_3.

A pair of elements a, b is *incomparable* if neither $a \prec b$ nor $b \prec a$. A sequence of pairwise incomparable elements is called an *antichain*. The Graph Minor Theorem established that the set of graphs $\mathcal{G}$ with the minor relation has no infinite antichain. Thus for every infinite set of graphs, one of its members is isomorphic to a minor of another. This is the main result in "Graph Minors XX" (Robertson and Seymour, 2004). Diestel (2017) describes this theorem as "one which dwarfs any other result in graph theory and may doubtless be counted among the deepest theorems that mathematics has to offer."

Theorem 1.3.5. (*Graph Minor Theorem*) *For every infinite set of graphs, one of its members is isomorphic to a minor of another.*

To prove Theorem 1.3.5 the authors observed that if $\{G_1, G_2, \ldots\}$ is an infinite antichain, then none of $G_2, G_3, \ldots$ have a minor isomorphic to G_1. Therefore it suffices to prove that the class of graphs with no minor isomorphic to a specific graph H has no infinite antichain. To prove this statement, they developed a structure theorem for the class of graphs with no H-minor, first when H is planar (Robertson and Seymour, 1986), and subsequently when H is non-planar (Robertson and Seymour, 2003).

In Section 1.1 we described the Graph Minor Theorem as stating that "every infinite class of graphs closed under minors has a finite list of minimal excluded minors." This statement is equivalent to the statement in Theorem 1.3.5. To see this, observe that an infinite set of minimal excluded minors for a

---

21 A relation $\prec$ on a set S is a partial order if it is reflexive, antisymmetric, and transitive. The set S is called a partially ordered set or poset. Reflexive, antisymmetric, and transitive operations were defined in Section 1.2. For example, the set of natural numbers with the divisibility relation and the set of all subsets of a set with the usual subset relation are partially ordered sets.

minor-closed class would be an infinite antichain. Theorem 1.3.5 implies that every minor-closed class of graphs has a finite list of minimal excluded minors.

The converse is also fairly straightforward. Suppose every minor-closed class of graphs has a finite list of minimal excluded minors. Let $\mathcal{H}$ be any set of graphs and consider the class of graphs with no minor isomorphic to any graph in $\mathcal{H}$. Since this class is closed under minors, it has a finite list of minimal excluded minors $G_1, \ldots, G_t$. By definition $G_1, \ldots, G_t$ are in $\mathcal{H}$, but no proper minor of $G_1, \ldots, G_t$ is in $\mathcal{H}$. This means that every graph in $\mathcal{H}$ has a minor isomorphic to at least one of $G_1, \ldots, G_t$. Therefore $\mathcal{H}$ has at least two graphs such that one is a minor of the other.

A *quasi order* is a relation that is reflexive and transitive. It is not necessarily antisymmetric. A quasi order is a *well-quasi ordering* if every sequence of elements $x_1, x_2, \ldots$ contains an increasing pair $x_i \prec x_j$, where $i < j$. So a well-quasi ordering has no infinite antichain nor an infinite strictly decreasing sequence. Clearly the class of graphs $\mathcal{G}$ with the minor relation has no infinite strictly decreasing sequence because deleting and contracting edges in a graph eventually results in the empty graph. The hard part is to show that $\mathcal{G}$ has no infinite antichain, which is Theorem 1.3.5. The Graph Minor Theorem is often presented as the following statement: The class of graphs $\mathcal{G}$ under the minor relation is well-quasi ordered.

Exercises

1.1 Suppose a graph has 15 edges, 3 vertices of degree 4, and all other vertices have degree 3. How many vertices does the graph have?

1.2 Is it possible to have a group of nine people where each person is acquainted with exactly five of the other people? Either draw a graph that demonstrates this or explain why it is not possible.

1.3 Prove that a graph on $n \geq 2$ vertices has at least two vertices of the same degree.

1.4 Figure 1.31 displays six graphs and three digraphs.
(a) For each graph find the degree sequence, mean degree, and density.
(b) For each digraph find the indegree sequence, outdegree sequence, mean indegree, mean outdegree, and density.

1.5 Draw all the non-isomorphic connected graphs on 5 vertices.[22]

---

22 The On-Line Encyclopedia of Integer Sequences (http://oeis.org) has the number of non-isomorphic connected graphs. Check to see the largest value of n for which these numbers are known.

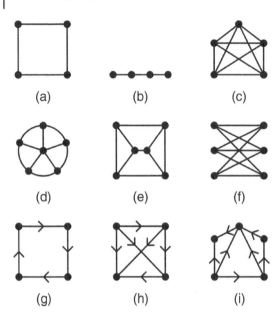

Figure 1.31 Examples of graphs and digraphs.

(a)   (b)   (c)

(d)   (e)   (f)

(g)   (h)   (i)

1.6   Draw all the non-identical graphs on 4 vertices and 3 edges.

1.7   How many labeled isomorphic graphs are there on 10 vertices? How many of them have 15 edges?

1.8   Are the pairs of graphs in Figure 1.32 isomorphic? If yes, give the isomorphism. If no, find an invariant that differs for them.

1.9   Give an example of two non-isomorphic graphs with the same degree sequence.

1.10   Show that the Petersen graph has a $K_{3,3}$-minor.

1.11   For the graphs in Figure 1.31:
(a) Find the decks of vertex-deletions and edge-deletions.
(b) Find minors isomorphic to W_3 and W_4, if possible.
(c) Find an induced subgraph isomorphic to $K_{1,3}$, if possible.
(d) Find their complements.
(e) Find their line graphs.

1.12   Illustrate $P_3 + W_4$, $P_3 \times W_4$, and $W_4 \oplus_3 W_4$.

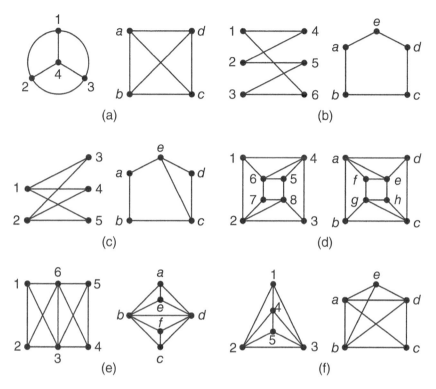

Figure 1.32 Pairs of graphs for isomorphism testing.

1.13 The *tensor product* of two graphs G and H, denoted by $G \otimes H$, has vertex set $V(G) \times V(H)$ and pairs (g, h) and (g, h') are adjacent if and only if gg' is an edge in G and hh' is an edge in H. Illustrate $P_4 \otimes P_4$, $P_2 \otimes C_3$, and $P_2 \otimes W_4$.

1.14 Show that each of the nine graphs in Figure 1.28 is not a line graph.

1.15 Figure 1.33 contains the nine graphs shown in Figure 1.28 as induced subgraphs. Find each of them.

Topics for Deeper Study

1.16 Give an example to show that digraphs are not reconstructable from their decks of vertex-deletions (Stockmeyer, 1977).

1.17 Line graphs have many equivalent characterizations and interesting properties. Let G be a connected graph.

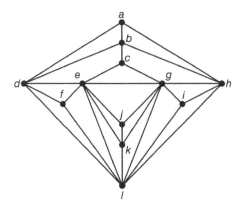

Figure 1.33 A graph that contains all nine forbidden induced subgraphs for line graphs.

(a) Prove that G is a line graph if and only if the edges of G can be partitioned into cliques so that every vertex is in at most two cliques (Krausz, 1943).

(b) What sort of graphs are isomorphic to their line graphs? For example, $L(C_n) \cong C_n$. As it turns out, cycles are the only graphs with this property. Let G be a connected graph with n vertices. Prove that $G \cong L(G)$ if and only if $G \cong C_n$.

(c) It is easy to check that $L(C_3) \cong L(K_{1,3})$ even though $C_3 \not\cong K_{1,3}$. Are there other such graphs? This was a question that Hassler Whitney answered in his paper introducing line graphs (Whitney, 1932a). Let G_1 and G_2 be connected graphs neither of which is C_3 or $K_{1,3}$. Prove that $G_1 \cong G_2$ if and only if $L(G_1) \cong L(G_2)$.

(d) Prove Theorem 1.3.3: A graph is a line graph if and only if it does not have any of the nine graphs in Figure 1.28 as induced subgraphs (Beineke, 1970).

(e) Suppose G is a graph with $\delta(G) \geq 5$. Prove that G is a line graph if and only if G has no induced subgraph isomorphic to the first six graphs in Figure 1.28 (Metelsky and Tyshkevich, 1997).

1.18 The set of automorphisms of a graph together with the composition operation form a group called the *automorphism group* of the graph. Explore the connection between graphs and groups (Chartrand et al., 2011).

1.19 Isomorphism and label isomorphism are not the only ways to determine if two graphs are similar. Ronald Graham notes that "Among a variety of fundamental themes running through Stan Ulam's mathematical research, one that particularly intrigued him was that of similarity (Graham, 1987). He was constantly fascinated by the problem of quantifying exactly how alike (or different) two mathematical objects or structures were." Ulam's idea for measuring the difference between k objects was to break them into as few as possible equal pieces. Let $\mathcal{G}_{n,m}$ be the set of graphs with m edges and at most n vertices An *Ulam decomposition* of k graphs $G_1, \ldots, G_k \in \mathcal{G}_{n,m}$ is k sets of partitions

$$(E_{11}, \ldots, E_{1t}), (E_{21}, \ldots, E_{2t}), \ldots, (E_{k1}, \ldots, E_{kt})$$

where, as graphs, $E_{1j} \cong E_{2j} \cong \ldots E_{kj}$ for $1 \leq j \leq t$. Let $U(G_1, \ldots, G_k)$ be the minimum value of t for which an Ulam decomposition of $G_1, \ldots, G_k$ exists. Let $U_k(n)$ be the maximum value of $U(G_1, \ldots, G_k)$, where the maximum is taken over all sets of k graphs from $\mathcal{G}_{n,m}$. This measure $U_k(n)$ is a measure of maximum dissimilarity for a set of k graphs in $\mathcal{G}_{n,m}$. The first result in the following text appears in Chung et al. (1979) and the second and third in Chung et al. (1981)

(a) Show that $U_2(n) \leq \frac{2}{3}n + c_2$, where c_2 is a constant.

(b) Show that $U_3(n) \leq \frac{3}{4}n + c_3$, where c_3 is a constant.

(c) Show that $U_k(n) \leq \frac{3}{4}n + c_k$, where c_k is a constant.

(The constant factor of $\frac{3}{4}$ did not increase for $k \geq 3$ suggesting that $\mathcal{G}_{n,m}$ is 3-dimensional in the sense that once there are three graphs that are maximally dissimilar, adding further graphs does not change anything.)

2

Fundamental Topics

In this chapter, we will tighten the precision of our concepts and develop them further. The focus is on crafting theorems. Section 2.1 presents a collection of results on trees that form a good introduction to induction arguments. Section 2.2 presents in detail the concept of distance in a graph. Section 2.3 presents a method of determining whether or not a sequence of numbers is the degree sequence of a graph. The results in Section 2.3 convey how good theorems can be obtained by making a few definitions and asking a simple question. This section is optional. Section 2.4 presents topics in spectral graph theory that are essential for the applications in Sections 3.3 and 3.4. The linear algebra prerequisites for Section 2.4 are in Appendix A.

2.1 Trees

Let G be a graph and let u and v be two vertices in G. Recall from Section 1.1 that u is connected to v if there is a $u - v$ path in G. A graph is connected if every pair of vertices is connected. Otherwise, it is disconnected. The relation "is connected to" is an equivalence relation on the set of vertices. To see this, observe that it is:

- Reflexive since a vertex is connected to itself.
- Symmetric since if there is a path from u to v, then there is a path from v to u.
- Transitive since if there is a path from u to v and a path from v to w, then there is a path from u to w.

This equivalence relation partitions the set of vertices into equivalence classes. Two vertices are in the same equivalence class if and only if there is a path from one of the vertices to the other. The subgraph induced by the vertices in an equivalence class is called a *component* of G (or connected component for extra clarity). A graph is connected if and only if it has exactly one component.

Graphs and Networks, First Edition. S. R. Kingan.
© 2022 John Wiley & Sons, Inc. Published 2022 by John Wiley & Sons, Inc.

$n = 1$ $n = 2$ $n = 3$ $n = 4$

Figure 2.1   Non-isomorphic trees on $n \leq 4$ vertices.

Recall from Section 1.1 that a connected acyclic graph is called a tree and an acyclic graph is called a forest. Figure 2.1 displays the non-isomorphic trees on at most 4 vertices.

Proposition 2.1.1.   *A graph is a tree if and only if every pair of distinct vertices is joined by a unique path.*

Proof. Suppose G is a tree and u and v are distinct vertices in G. Since G is connected, there is a path P_1 from u to v. Suppose, if possible, there is another path P_2 from u to v, possibly overlapping with P_1. Let x be the first vertex along P_1 after which P_1 and P_2 differ and let y be the next vertex on P_1 after which P_1 and P_2 overlap, as shown in the following diagram. Then there is a cycle in G consisting of path P_1 from x to y and path P_2 from y to x. This is a contradiction to the hypothesis. Therefore u are v are joined by a unique path.

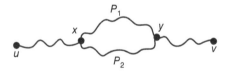

Conversely, suppose every pair of vertices in a graph G is joined by a unique path. Then clearly G is connected. Moreover, G cannot have a cycle, because if it did, there would be two distinct paths between every pair of vertices on the cycle. Therefore G is a tree.   □

Proposition 2.1.2.   *If G is a tree with n vertices and m edges, then $n = m + 1$.*

Proof. The proof is by induction on n. If $n = 1$, then G has one vertex and zero edges and the result holds. Assume the result holds for all trees with fewer than n vertices and let G be a tree on $n \geq 2$ vertices. Pick any edge e and delete it (deleting an edge leaves the end-vertices intact). Proposition 2.1.1 implies that $G \backslash e$ is a disconnected graph with exactly two components H_1 with n_1 vertices and m_1 edges and H_2 with n_2 vertices and m_2 edges. Both H_1 and H_2 are trees. By the induction hypothesis $n_1 = m_1 + 1$ and $n_2 = m_2 + 1$. Therefore

$$n = n_1 + n_2 = m_1 + 1 + m_2 + 1 = m_1 + m_2 + 1 + 1.$$

Since m is the sum of m_1 and m_2 and the edge e that was removed,

$$m = m_1 + m_2 + 1.$$

Therefore $n = m + 1$. $\square$

Corollary 2.1.3. *If G is a forest with $n \geq 2$ vertices, m edges, and t components, then $n = m + t$.*

Proof. Since G is a forest, it has connected components $G_1, \ldots, G_t$ each of which are trees with n_i vertices and m_i edges for $1 \leq i \leq t$. Proposition 2.1.2 implies that $n_i = m_i + 1$ for each G_i. Therefore

$$n = n_1 + \cdots + n_t = m_1 + 1 + \cdots + m_t + 1 = m + t.$$

$\square$

A vertex of degree 1 is called a *leaf*. Clearly, the path graph P_n, where $n \geq 2$, has exactly two leaves. The next proposition establishes that all trees have at least two leaves.

Proposition 2.1.4. *A tree on $n \geq 2$ vertices has at least two leaves.*

Proof. Let G be a tree with $n \geq 2$ vertices and $m \geq 1$ edges. Since G is connected, $deg(v) \geq 1$ for every vertex v. If G has no leaves, then each vertex has degree at least 2, and consequently there are two distinct paths from each vertex to every other vertex. This is a contradiction to Proposition 2.1.1. If G has exactly one leaf, then the leaf has degree 1 and the other $n - 1$ vertices have degree at least 2. So

$$\sum deg(v) \geq 1 + 2(n - 1) = 2n - 1.$$

However, by Propositions 1.1.1 and 2.1.2,

$$\sum deg(v) = 2m = 2(n - 1) = 2n - 2.$$

So we get $2n - 2 \geq 2n - 1$, which is impossible. Therefore G has at least two leaves.
$\square$

A subgraph T of a graph G is called a *spanning tree* if T is a tree and $V(T) = V(G)$. Figure 2.2 displays two examples of spanning trees (highlighted in bold). The next result is a necessary and sufficient condition for a connected graph in terms of spanning trees.

Figure 2.2   Spanning trees.

Proposition 2.1.5.   *A graph is connected if and only if it has a spanning tree.*

Proof. Suppose G is a connected graph. Then G has a connected spanning subgraph H. If H has no cycles, then H is the required spanning tree. If H has a cycle, then removing any edge of this cycle leaves a connected subgraph. If this subgraph is a tree, we are done, otherwise continue in this manner until no cycles remain. The converse follows from the definition of a spanning tree.   □

Proposition 2.1.6.   *Suppose G has n vertices and m edges. The following statements are equivalent:*

  (i) *G is a tree;*
 (ii) *G is connected and $n = m + 1$; and*
(iii) *G has no cycles and $n = m + 1$.*

Proof. Suppose G is a tree. By definition G is connected and Proposition 2.1.2 implies that $n = m + 1$. Therefore (i) implies (ii).

Next, suppose G is connected and $n = m + 1$. We must prove that G has no cycles. Since G is connected, G has a spanning tree by Proposition 2.1.5. The spanning tree has n vertices and $n - 1$ edges by Proposition 2.1.2. However, G also has $n - 1$ edges by the hypothesis. So the spanning tree must be G itself. Therefore G has no cycles.

Lastly, suppose G has no cycles and $n = m + 1$. We must prove that G is a tree. Let $G_1, \ldots, G_t$ be the connected components of G and let each G_i have n_i vertices and m_i edges where $i \in \{1, \ldots, t\}$. Since G has no cycles, each G_i has no cycles, and is therefore a tree. Proposition 2.1.2 implies that $n_i = m_i + 1$. So

$$n = m + 1 = (m_1 + m_2 + \cdots + m_t) + 1 = n_1 - 1 + \cdots n_t - 1 + 1 = n - t + 1.$$

Since equality holds throughout, $t = 1$. Therefore G is connected, and a connected graph with no cycles is a tree.   □

We will end this section by discussing the origins of trees. As mentioned in Section 1.1, Cayley was trying to enumerate the isomers of saturated hydrocarbons (chemicals of the form C_kH_{2k+2} shown in Figure 1.10). He proved that graphs corresponding to $C_k H_{2k+2}$ form trees with vertices of degree 1 or 4. This is because the number of vertices in $C_k H_{2k+2}$ is

$$k + 2k + 2 = 3k + 2,$$

and since each vertex labeled C has degree 4 and each vertex labeled H has degree 1,

$$\sum deg(v) = 4k + 1(2k + 2) = 6k + 2 = 2(3k + 1).$$

Proposition 1.1.1 implies that the number of edges is $3k + 1$. Since the graph is connected and the number of edges is one less than the number of vertices, the graph is a tree by Proposition 2.1.6(ii).

Thus the problem of counting the isomers of the saturated hydrocarbons is the same as the problem of counting all trees in which every vertex has degree 1 or 4. Cayley calculated graphs of the form $C_k H_{2k+2}$ up to $k = 6$, and in this manner predicted the existence of unknown isomers. Subsequently they were found and Cayley's predictions were correct (Thorpe, 1910).[1]

Cayley also gave a formula for the number of labeled trees on n vertices (Cayley, 1889). He proved that the number of labeled trees on n vertices is n^{n-2}. Cayley credits the formula to Borchardt (1861) in his paper. We present André Joyal's proof (Joyal, 1981) taken from Aigner and Ziegler (2010).

Joyal's proof sets up a bijection between a set with n^n elements and the set of trees with n vertices together with two vertices called the left vertex L and the right vertex R, which could be the same. For example, let $X_9 = \{1, 2, \ldots, 9\}$. Consider the set of all functions from X_9 to X_9 and denote this set as $\mathcal{X}_9$. Observe that there are 9^9 such functions, so $|\mathcal{X}_9| = 9^9$. An example of a function from X_9 to X_9 is

$$f = \begin{pmatrix} 1 & 2 & 3 & 4 & 5 & 6 & 7 & 8 & 9 \\ 4 & 4 & 5 & 6 & 3 & 1 & 2 & 8 & 5 \end{pmatrix}.$$

A digraph, denoted by $\vec{G}_f$, can be associated with f in a natural way by drawing vertices labeled $1, \ldots, 9$ and directed arcs from i to j if $f(i) = j$, as shown in the first diagram of Figure 2.3. Observe that vertices 1, 3, 4, 5, 6, and 8 are in cycles and 2, 7, and 9 are not in cycles. Let $Y = \{1, 3, 4, 5, 6, 8\}$. Then

$$f|_Y = \begin{pmatrix} 1 & 3 & 4 & 5 & 6 & 8 \\ 4 & 5 & 6 & 3 & 1 & 8 \end{pmatrix}.$$

This gives an ordering 4, 5, 6, 3, 1, 8 according to the second row of $f|_Y$, with a left end vertex 4 and a right end vertex 8. A tree T can be constructed as follows: Draw 4, 5, 6, 3, 1, 8 in this order as a path from vertex 4 to 8 and fill in the remaining vertices as in $\vec{G}_f$ without arrows. This tree T with the path from 4 to 8 highlighted is shown in the second diagram in Figure 2.3.

---

1 For a more recent survey see "Enumerating Molecules" that describes how the difficulties that arose in using Cayley's methods led to Pólya's Theory of Counting (Faulon et al., 2005).

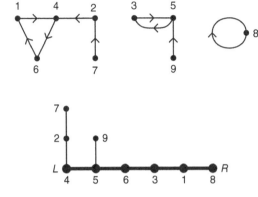

Figure 2.3 An example for Cayley's tree counting theorem.

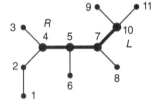

This procedure can be reversed. Consider the tree shown in Figure 2.4 with two flagged vertices $L = 10$ and $R = 4$, then the path $10, 7, 5, 4$ determines that $Y = \{4, 5, 7, 10\}$. Writing these values in ascending order on top and in the order of occurrence in the path on the bottom, from the vertex marked L to the vertex marked R, we have

$$f|_Y = \begin{pmatrix} 4 & 5 & 7 & 10 \\ 10 & 7 & 5 & 4 \end{pmatrix}.$$

Figure 2.4 A tree with a left and right vertex.

For each vertex v not in Y, there is a unique path in the tree from v to a vertex in Y, and this path does not contain any other vertices in Y. For example, the unique path from $1 \notin Y$ to $4 \in Y$ is $1, 2, 4$. So in the function f to be constructed 1 is mapped to 2 and 2 is mapped to 4. Thus the function corresponding to this tree is given by

$$f = \begin{pmatrix} 1 & 2 & 3 & 4 & 5 & 6 & 7 & 8 & 9 & 10 & 11 \\ 2 & 4 & 4 & 10 & 7 & 5 & 5 & 7 & 10 & 4 & 10 \end{pmatrix}.$$

This procedure can be generalized to establish a formula for the number of labeled trees on n vertices.

Theorem 2.1.7. (Cayley's Tree Counting Theorem) *There are n^{n-2} distinct labeled trees with $n \geq 2$ vertices.*

Proof. Let T_n be the number of distinct labeled trees on n vertices where $n \geq 2$. Let $X_n = \{1, 2, \ldots, n\}$ be the set of natural numbers between 1 and n both inclusive. We denote by $\mathcal{X}_n$ the set of functions $f : X_n \to X_n$ given by

$$\begin{pmatrix} 1 & 2 & \cdots & n \\ f(1) & f(2) & \cdots & f(n) \end{pmatrix}.$$

Observe that $|\mathcal{X}_n| = n^n$.

Given a tree T with n vertices, designate one vertex as the "left" vertex L and one vertex as the "right" vertex R. The left and right vertices could be the same. Let $\mathcal{T}_n$ be the set of all such combinations (T, L, R) for all trees T with n vertices. There are n choices for L and n choices for R, so $|\mathcal{T}_n| = n^2 T_n$. To prove the theorem we will prove that $|\mathcal{T}_n| = n^n$ by establishing a bijection between $\mathcal{T}_n$ and $\mathcal{X}_n$.

Given a function $f : X_n \to X_n$ we may construct a digraph $\vec{G}_f$ with n vertices labeled $1, 2, \ldots, n$ by drawing an arc from i to $f(i)$ for each $1 \leq i \leq n$. In this digraph every vertex has outdegree 1 and the number of edges is equal to the number of vertices since each number i has only one corresponding value $f(i)$.

Consider the components of $\vec{G}_f$ and observe that the number of vertices and edges in each component is equal. So a component cannot be a tree and must contain at least one cycle. However, since each vertex has outdegree 1, there can be only one cycle in a component. Let Y be the subset of vertices of the cycles of $\vec{G}_f$, and denote the elements of Y in ascending order as $i_1, i_2, \ldots, i_k$. We may express $f|_Y$ as

$$\begin{pmatrix} i_1 & i_2 & \cdots & i_k \\ f(i_1) & f(i_2) & \cdots & f(i_k) \end{pmatrix}.$$

Observe that Y is the unique largest subset of X_n such that $f|_Y$ is a permutation. To see this, let $S \subseteq X_n$ and let ϕ be a permutation of S. Since $\phi(i) \in S$ for every $i \in S$, if we begin with one value $i \in S$ and evaluate $\phi(i)$, then $\phi(\phi(i))$, then $\phi(\phi(\phi(i)))$, and so on, eventually we must reach a point where the value is i again. This corresponds to a cycle in $\vec{G}_f$, so S must be a union of cycles in $\vec{G}_f$ and is therefore a subset of Y.

We can construct a tree with n vertices and a designated left vertex and right vertex in the following manner. First construct a path joining vertices $f(i_1), f(i_2), \ldots, f(i_k)$. Designate $f(i_1)$ to be the left vertex L and $f(i_k)$ to be the right vertex R. Then create vertices for all values in $X_n - Y$. There is a unique path in $\vec{G}_f$ from each such vertex to a vertex in Y not including any other vertices in Y. Connect these vertices to the ones in Y using undirected edges corresponding to the directed edges in these paths.

Since each step in the procedure is determined by the function f, the procedure describes a function from $\mathcal{X}_n$ to $\mathcal{T}_n$. Furthermore, any change in the function f will result in a different tree and a different designation of L and R; therefore the function is a one-to-one function.

Finally, if we begin with a tree T with n vertices and a designation of left and right vertices, L and R, then since there is a unique path from L to R in T, we may construct a set Y using the vertices in that path in order, and reverse the procedure to obtain a corresponding function $f \in \mathcal{X}_n$. Therefore the procedure describes a function that is onto, which is therefore a bijection. So $n^n = |\mathcal{X}_n| = |\mathcal{T}_n| = n^2 T_n$ and consequently $T_n = n^{n-2}$. □

2.2 Distance

Recall from Section 1.1 that the distance between two vertices u and v in a connected graph, denoted by $d(u, v)$, is the length of the shortest path between u and v. If the graph is disconnected, then the distance between two vertices is viewed as infinite. Distance is a metric[2] on the vertex set of a graph since it satisfies the four properties of a metric:

- The distance between u and v is greater than or equal to zero;
- The distance between u and v is zero if and only if $u = v$;
- The distance between u and v is the same as the distance between v and u; and
- Given three vertices u, v, w, we can get from u to v by the shortest path or make a detour via w (assuming w is not on the shortest path) and therefore

$$d(u, v) \leq d(u, w) + d(w, v).$$

Let u be a vertex of a connected graph G. The *eccentricity* of u, denoted by $ecc(u)$, is the maximum distance from u to any other vertex in G. Consider the graph in Figure 2.5. Observe that:

$$ecc(a) = ecc(h) = 4, ecc(b) = ecc(c) = ecc(d) = ecc(e) = ecc(g) = 3,$$

$$ecc(f) = 2.$$

In Section 1.1 the diameter of a connected graph was defined as the maximum distance between a pair of vertices. In terms of eccentricity,

$$diam(G) = max\{ecc(u) \mid u \in V(G)\}.$$

---

2 Let S be a set and d be a real-valued function on $S \times S$. Then d is called a *metric* if it satisfies four properties:

(1) $d(x, y) \geq 0$ for all x, y in S;
(2) $d(x, y) = 0$ if and only if $x = y$;
(3) $d(x, y) = d(y, x)$ for all x, y in S; and
(4) $d(x, y) \leq d(x, z) + d(y, z)$ for all x, y, z in S.

The set S with metric d is called a *metric space*. The last property is the triangle inequality. For example, the usual distance function $d(x, y) = |x - y|$ is a metric on the set of real numbers.

Figure 2.5 Eccentricity, diameter, and radius.

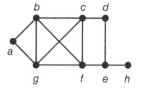

The *radius* of G, denoted by $rad(G)$, is defined as

$$rad(G) = min\{ecc(u) \mid u \in V(G)\}.$$

The graph in Figure 2.5 has $diam(G) = 4$ and $rad(G) = 2$. For K_n, where $n \geq 2$,

$$rad(K_n) = diam(K_n) = 1.$$

For $K_{r,s}$, where $r, s \geq 2$,

$$rad(K_{r,s}) = diam(K_{r,s}) = 2.$$

For the star graph with n vertices, $K_{1,n-1}$, $rad(K_{1,n-1}) = 1$ and $diam(K_{1,n-1}) = 2$.

The notation $\lceil x \rceil$ stands for the smallest integer greater than or equal to x and $\lfloor x \rfloor$ stands for the greatest integer less than or equal to x. For P_n, where $n \geq 2$,

$$diam(P_n) = n - 1 \text{ and } rad(P_n) = \left\lfloor \frac{n}{2} \right\rfloor.$$

For C_n, where $n \geq 3$,

$$rad(C_n) = diam(C_n) = \left\lfloor \frac{n}{2} \right\rfloor.$$

There are no precise formulas for computing radius and diameter in general; the best we can do is find bounds for them.

Proposition 2.2.1. *Let G be a connected graph. Then, $rad(G) \leq diam(G) \leq 2rad(G)$.*

Proof. Clearly $rad(G) \leq diam(G)$ since diameter is the maximum distance between two vertices and radius is the minimum distance between two vertices. We must prove that $diam(G) \leq 2rad(G)$. Let u and v be two vertices such that $d(u, v) = diam(G)$. Let w be a vertex of minimum eccentricity. Then

$$rad(G) = ecc(w) = max\{d(w, z) \mid z \in V(G)\}.$$

By the Triangle Inequality,

$$d(u, v) \leq d(u, w) + d(w, v) \leq rad(G) + rad(G) = 2rad(G).$$

Therefore $diam(G) \leq 2rad(G)$. □

A vertex whose eccentricity is equal to the radius is called a *central vertex*. The subgraph induced by the central vertices is called the *center* of G. A vertex whose eccentricity is equal to the diameter is called a *peripheral vertex*. The subgraph induced by the peripheral vertices is called the *periphery* of G. For the graph in Figure 2.5, the girth is 3, f is the only central vertex, and a and h are peripheral vertices. Consequently, the center of this graph is the single vertex $\{f\}$ and the periphery is the set of two isolated vertices $\{a, h\}$.

Let v be a vertex in a connected graph. The *total distance*[3] of v, denoted by $td(v)$, is defined as

$$td(v) = \sum_{w \in V} d(v, w).$$

For example, in K_n, $td(v) = 1(n - 1) = n - 1$ since every vertex is distance one from the other $n - 1$ vertices.

Consider the star graph $K_{1,n-1}$ with vertex u in one class and vertices $\{v_1, \ldots, v_{n-1}\}$ in the other class. Vertex u is distance one from the other $n - 1$ vertices, so $td(u) = n - 1$. Vertices $v_1, \ldots, v_{n-1}$ are distance two from each other. So for $1 \leq i \leq n - 1$,

$$td(v_i) = 1 + 2(n - 2) = 2n - 3.$$

A vertex v is called a *median vertex* if it has the smallest total distance among all the vertices. The *median* of a graph is the subgraph induced by its median vertices. For the graph in Figure 2.5, $td(c) = td(f) = 10$ and these two vertices have the smallest total distance. Thus median vertices are c and f and the median subgraph is the edge cf.

Let G be a connected graph with at least 3 vertices. A vertex whose removal disconnects G is called a *cut vertex*. An edge whose removal disconnects G is called a *bridge*. Consider the connected graphs G_1 and G_2 in Figure 2.6. In the first graph, edge e is a bridge since its removal disconnects the graph. Both end vertices of edge e (vertices v_5 and v_6) are cut vertices. In the second graph removing edge e disconnects the graph because of the isolated vertex v_6 that is left behind. Therefore e is a bridge. Observe that v_4 is a cut vertex, but the other end vertex v_6 is not a cut vertex.

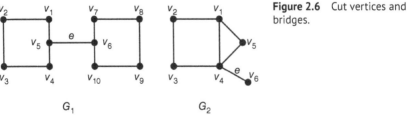

Figure 2.6 Cut vertices and bridges.

G_1 G_2

---

3 Total distance of a vertex is called *status* of a vertex in Buckley and Harary (1990) and Harary (1959). In social network analysis it is called *reach* of a vertex.

A *block* in a connected graph is a maximal induced subgraph with no cut vertices. Here maximal means that the induced subgraph has as many vertices as possible provided no vertex is a cut vertex. For example, consider the first graph in Figure 2.6. It has three blocks B_1, B_2, and B_3, where B_1 is the subgraph induced by $\{v_1, v_2, v_3, v_4, v_5\}$, B_2 is the subgraph induced by $\{v_5, v_6\}$, and B_3 is the subgraph induced by $\{v_6, v_7, v_8, v_9, v_{10}\}$. The second graph has two blocks B_1 and B_2, where B_1 is the subgraph induced by $\{v_1, v_2, v_3, v_4, v_5\}$ and B_2 is the subgraph induced by $\{v_4, v_6\}$

Observe that in a path on $n \geq 3$ vertices, every vertex except the two end vertices is a cut vertex and every edge is a bridge. In particular P_n has two vertices that are not cut vertices. The next proposition establishes that this is true for all graphs. The proof is taken from Chartrand et al. (2011).

Proposition 2.2.2. *Let G be a connected graph with $n \geq 3$ vertices. Then G has at least two vertices that are not cut vertices.*

Proof. Suppose, if possible, G has at most one vertex that is not a cut vertex. Let u and v be vertices of G such that $d(u, v) = diam(G)$. By the assumption, one of u or v must be a cut vertex. Suppose v is a cut vertex. Consider the graph $G - v$ and let w be a vertex in the component of $G - v$ that does not contain u. Since u and w are in different components of $G - v$, every path from u to w must contain v. Therefore $d(u, w) > d(u, v) = diam(G)$. This is a contradiction since diameter is the maximum distance. Thus G has at least two vertices that are not cut vertices. □

A cycle with an even number of vertices is called an *even cycle*, otherwise it is called an *odd cycle*. In the next theorem we will prove that a graph is bipartite if and only if it has no odd cycles. First we need a lemma describing a property of shortest paths, referred to as the "optimal substructure" property of shortest paths.

Lemma 2.2.3. *Let G be a connected graph. A shortest path from vertex s to vertex t contains within it a shortest path from s to any other vertex in the path before t.*

Proof. Suppose an $s - t$ path P given by $s v_1 v_2 \cdots v_{k-1} t$ is a shortest path from s to t. Then for every v_i along this path, the portion from s to v_i is a shortest $s - v_i$ path, because otherwise there would be a shorter path P' from s to v_i, which could be extended to t along P to give a shorter $s - t$ path; a contradiction. □

Theorem 2.2.4. *Let G be a connected graph. Then G is bipartite if and only if every cycle is an even cycle.*

Proof. Suppose G is bipartite. Then the vertex set V is the union of two disjoint sets V_1 and V_2 such that the edges of G join vertices in V_1 to vertices in V_2. Let C

be a cycle in G. The vertices of C must alternate between V_1 and V_2. Therefore C must be an even cycle.

Conversely, suppose every cycle in G has even length. For each vertex $v \in V(G)$ define two sets related to v:

$$v_{even} = \{w \in V \mid d(v, w) \text{ is even}\}$$

and

$$v_{odd} = \{w \in V \mid d(v, w) \text{ is odd}\}.$$

Observe that v_{even} and v_{odd} are disjoint and $V = v_{even} \cup v_{odd}$. We will prove that there are no edges between vertices in v_{even}.

Suppose, if possible, there is an edge between two vertices x and y in v_{even}. Let P_1 and P_2 be the shortest paths from v to x and v to y, respectively. Both paths P_1 and P_2 have even length. If P_1 and P_2 do not overlap, then P_1, P_2, and the edge xy forms a cycle of odd length as shown in the following diagram. This is a contradiction to the hypothesis.

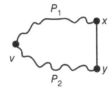

Otherwise, suppose P_1 and P_2 overlap and z is the last common vertex as shown in the following diagram. We will prove that paths $z - x$ along P_1 and $z - y$ along P_2 both have odd length or both have even length. Suppose this is not the case and assume that $z - x$ has odd length and $z - y$ has even length. Since P_1 and P_2 have even length, the path $v - z$ along P_1 has odd length and the path $v - z$ along P_2 has even length and by Lemma 2.2.3 one of these is not the shortest path from v to z. Consequently either P_1 is not the shortest path from v to x or P_2 is not the shortest path from v to y; a contradiction.

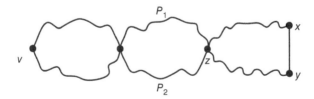

Thus $z - x$ along P_1 and $z - y$ along P_2 both have odd length or both have even length. So, the cycle formed with these two paths and edge xy has odd length; a contradiction to the hypothesis.

Thus we conclude that there is no edge between vertices in v_{even}. Similarly, there is no edge between vertices in v_{odd}. Therefore the graph is bipartite. ☐

The *mean distance* in a connected graph G is given by

$$\mu(G) = \frac{1}{\binom{n}{2}} \sum_{u,v \in V} d(u, v),$$

where the sum is taken over every pair of vertices in G. Alternatively, $\mu(G)$ may be calculated by adding the total distance of each vertex and dividing by $\binom{n}{2}$, and then halving the number since each vertex is counted twice. The definition of mean distance appears in Doyle and Graver (1977), but the concept dates back to 1947 when German chemist Harry Weiner observed that the melting point of certain hydrocarbons is directly proportional to

$$W(G) = \sum_{u,v \in V} d(u, v),$$

where G is the graph corresponding to the hydrocarbon (Wiener, 1947). In chemistry $W(G)$ is called the *Weiner index* of the hydrocarbon. For example, consider the star graph $K_{1,n-1}$ with vertex u in one class and vertices $\{v_1, \ldots, v_{n-1}\}$ in the other class. As mentioned earlier, vertex u is distance one from the other $n-1$ vertices, and vertices $v_1, \ldots, v_{n-1}$ are distance two from each other. So

$$W(K_{1,n}) = (n-1) + 2\binom{n-1}{2} = (n-1)^2,$$

and therefore

$$\mu(K_{1,n}) = \frac{(n-1)^2}{\binom{n}{2}} = \frac{2(n-1)}{n}.$$

For the graph in Figure 2.5, $W(G) = 51$ and the mean distance is

$$\mu(G) = \frac{51}{\binom{8}{2}} \approx 1.82.$$

For digraphs, the mean distance is defined in a similar manner provided the digraph is strongly connected. The denominator is, however, $n(n-1)$. Thus if D is a strongly connected digraph, then the mean distance is

$$\mu(D) = \frac{1}{n(n-1)} \sum_{u,v \in V} d(u, v).$$

The next theorem gives a sharp bound on $\mu(G)$ in terms of the number of vertices. The proof is taken from West (2001). We begin with a lemma.

Lemma 2.2.5. $\mu(P_n) = \frac{n+1}{3}$.

Proof. We will first prove that $W(P_n) = \binom{n+1}{3}$, where $n \geq 2$. This portion of the proof is by induction on the number of vertices. The result is true for $n = 2$ vertices since $W(P_2) = 1 = \binom{2+1}{3}$. Suppose the result is true for a path on $n - 1$ vertices. Let P_n be a path with vertices labeled in order as $v_1, \ldots, v_n$. We can think of P_n as P_{n-1} with a vertex v_n added to the end. Thus

$$W(P_n) = W(P_{n-1}) + td(v_n).$$

By the induction hypothesis $W(P_{n-1}) = \binom{n}{3}$ and

$$td(v_n) = 1 + 2 + \cdots + (n - 1) = \binom{n}{2}.$$

So

$$W(P_n) = \binom{n}{3} + \binom{n}{2}$$
$$= \frac{n(n - 1)(n - 2)}{6} + \frac{n(n - 1)}{2}$$
$$= \frac{n(n - 1)(n + 1)}{6}$$
$$= \binom{n + 1}{3}.$$

Therefore

$$\mu(P_n) = \frac{\binom{n+1}{3}}{\binom{n}{2}} = \frac{n + 1}{3}.$$

$\square$

Theorem 2.2.6. *Let G be a connected graph on $n \geq 2$ vertices. Then $\mu(G) \leq \frac{n+1}{3}$. Moreover, equality holds when G is isomorphic to P_n.*

Proof. Suppose T is a spanning tree for G. Since every $u - v$ path in T also appears in G, the shortest $u - v$ path in G is no longer than the shortest $u - v$ path in T. Therefore $W(G) \leq W(T)$. Next, we will prove by induction on the number of vertices that for any tree T on n vertices, $W(T) \leq W(P_n)$. The result is true for $n = 2$ vertices since P_2 is the only tree on 2 vertices. Suppose the result is true for trees with $n - 1$ vertices. Let u be a leaf of T. Then

$$W(T) = W(T - u) + td(u).$$

By the induction hypothesis $W(T - u) \leq W(P_{n-1})$. Therefore

$$W(T) \leq W(P_{n-1}) + td(u).$$

We will prove that $td(u)$ is maximized when T is the path P_n and u is an end vertex of P_n. Let k be the largest distance from u to another vertex v in the tree T. Since T is connected, there is a path from u to v. For each value $1 \leq j \leq k$, there is a vertex w along the path such that $d(u, w) = j$. The sum of the $n - 1$ values used to calculate $td(u)$ must contain all values from 1 to k. Observe that this sum is maximized when $k = n - 1$ and

$$td(u) = 1 + 2 + \cdots + (n - 1)$$

since any repeated values would reduce the total. Therefore $td(u)$ is maximized when T is the path P_n and u is an end vertex and $W(T) \leq W(P_n)$. By Lemma 2.2.5

$$W(G) \leq W(P_n) = \binom{n + 1}{3}.$$

$\square$

We will end this section with a sophisticated bound on the diameter in terms of the minimum degree (Erdős et al., 1989). The set $N(v) \cup \{v\}$ is called a *closed neighborhood*. This proof is one from the Book.[4]

Theorem 2.2.7. *If G is a connected graph with n vertices and minimum degree $\delta \geq 2$, then*

$$diam(G) \leq \left\lfloor \frac{3n}{\delta + 1} \right\rfloor.$$

Proof. Without loss of generality, suppose G has the maximum number of edges so that addition of one more edge reduces the diameter. Let $diam(G) = d$ and let x and y be a pair of vertices such that $d(x, y) = d$. For $0 \leq i \leq d$, define

$$S_i = \{v \in V(G) \mid d(x, v) = i\}.$$

Note that $|S_0| = 1$ and $|S_d| = 1$. (Think of sets S_i as concentric balls of vertices.) Since $deg(x) \geq \delta \geq 2$,

$$|S_{i-1}| + |S_i| + |S_{i+1}| \geq \delta + 1,$$

where $S_{-1} = S_{d+1} = \phi$. So

$$n = \sum |S_i| \geq \left\lfloor \frac{d}{3} \right\rfloor (\delta + 1),$$

and therefore

$$d \leq \left\lfloor \frac{3n}{\delta + 1} \right\rfloor.$$

$\square$

---

4 "The Book" is where God keeps the most elegant proof of every mathematical theorem, according to Paul Erdős. Occasionally we are allowed a glimpse inside the Book. Erdős famously said "You don't have to believe in God, but you should believe in the Book."

2.3 Degree Sequences

In this section we ask when an arbitrary sequence of non-negative integers $d_1, d_2, \ldots, d_n$ corresponds to the degree sequence of a graph. A sequence of non-negative integers $d_1, d_2, \ldots, d_n$ is called a *graphical sequence* if there is a graph with n vertices that has degree sequence $d_1, d_2, \ldots, d_n$. Since the maximum possible degree is $n-1$ and the sum of the degrees is even by Proposition 1.1.1, the following two conditions are necessary:

(1) $0 \le d_i \le n-1$ for $i \in \{1, \ldots, n\}$; and

(2) $\sum_{i=1}^{n} d_i$ is even.

However, these conditions are not sufficient. For example, consider the sequence of numbers

$$3, 3, 3, 1$$

as a possible candidate for a graph with 4 vertices. Clearly, this sequence meets the aforementioned two criteria, but there is no graph on 4 vertices that has this particular degree sequence. See Figure 1.16, which lists all the non-isomorphic graphs on 4 vertices.

The next result gives a necessary and sufficient condition for a sequence of non-negative integers to be a graphical sequence (Havel, 1955; Hakimi, 1962).

Theorem 2.3.1. (Havel–Hakimi Theorem) *Let* $d_1 \ge d_2 \ge \cdots \ge d_n$ *be a sequence of non-negative integers, where* $d_1 \ge 1$ *and* $n \ge 2$. *Then* $d_1, d_2, \ldots, d_n$ *is a graphical sequence if and only if* $d_2 - 1, d_3 - 1, \ldots, d_{d_1+1} - 1, d_{d_1+2}, \ldots, d_n$ *is a graphical sequence.*

Proof. Suppose $d_1, d_2, \ldots, d_n$ is a graphical sequence. Then there are graphs with vertices $v_1, v_2, \ldots, v_n$ having degrees $d_1, d_2, \ldots, d_n$, respectively. Among all such graphs choose a graph G in such a way that the sum of the degrees of the neighbors of v_1 is maximum. We will prove first that v_1 is adjacent to vertices having degrees $d_2, d_3, \ldots, d_{d_1+1}$.

Suppose this is not the case. Then there exists vertices v_r and v_s with $deg(v_r) > deg(v_s)$ such that v_1 is adjacent to v_s, but not to v_r. Since $deg(v_r) > deg(v_s)$, there exists a vertex v_t such that v_t is adjacent to v_r, but not to v_s. Remove edges v_1v_s and v_rv_t and add edges v_1v_r and v_tv_s as shown in the following diagram.

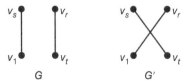

$$G \qquad\qquad G'$$

This results in a new graph G' having the same degree sequence as G. However, in G' the sum of the degrees of the neighbors of v_1 is more than in G since $deg(v_r) > deg(v_s)$. This is a contradiction to the maximality of G. Therefore we may assume v_1 is adjacent to vertices having degrees $d_2, d_3, \ldots, d_{d_1+1}$. Consequently when we remove vertex v_1, we get the graph $G - v_1$, which has degree sequence $d_2 - 1, d_3 - 1, \ldots, d_{d_1+1} - 1, d_{d_1+2}, \ldots, d_n$.

Conversely, suppose $d_2 - 1, d_3 - 1, \ldots, d_{d_1+1} - 1, d_{d_1+2}, \ldots, d_n$ is a graphical sequence and let G be a graph realizing this sequence. A new graph G' may be constructed by adding a new vertex v_1 and d_1 edges incident to V_1 and the vertices with degree sequence $d_2 - 1, d_3 - 1, \ldots, d_{d_1+1} - 1$. This graph G' has degree sequence $d_1, d_2, \ldots, d_n$, where $d_1 \geq d_2 \geq \cdots \geq d_n$. □

For example, consider the sequence

$$S : 5, 4, 4, 4, 3, 1, 1.$$

First observe that this sequence of numbers is a candidate for the degree sequence of a graph with 7 vertices and that it meets the two necessary conditions since $d_i \leq 6$ and $\sum d_i$ is even. Theorem 2.3.1 implies that we must remove d_1 and reduce the next five numbers by 1, to get the following sequence:

$$S_1 : 3, 3, 3, 2, 0, 1.$$

Rewrite the sequence so that the numbers are in descending order:

$$S_1 : 3, 3, 3, 2, 1, 0.$$

Again, this sequence meets the two necessary conditions for being a graphical sequence. Therefore we may remove d_1 and reduce the next three numbers by 1 to get:

$$S_2 : 2, 2, 1, 1, 0.$$

Proceeding in this manner gives

$$S_3 : 1, 0, 1, 0,$$

which can be rewritten as

$$S_3 : 1, 1, 0, 0,$$

and finally

$$S_4 : 0, 0, 0.$$

When the sequence of zeros is reached, we may stop and conclude the sequence is graphical.

This process can be reversed to give a graph with degree sequence S. Start with S_3 and join two vertices as shown in the following diagram. Add a new vertex and join it to the previous ones so as to obtain a graph with degree sequence S_3. Repeat to obtain S_2, and repeat once more to obtain S. The final graph has degree sequence S.

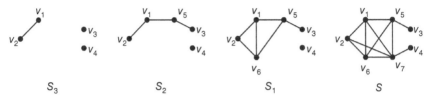

$$S_3 \qquad\qquad S_2 \qquad\qquad S_1 \qquad\qquad S$$

Consider another example of a sequence

$$S : 4, 4, 3, 2, 1.$$

This sequence meets the two necessary conditions for being graphical. Now, if it is graphical, the graph would have 5 vertices. Since $d_1 = 4$, remove d_1 and reduce the next four vertices by 1 to get:

$$S_1 : 3, 2, 1, 0.$$

If this sequence is graphical, the graph would have 3 vertices, so d_i can be at most 2. Since s_1 does not meet the first necessary condition, it is not graphical, and therefore neither is s.

Theorem 2.3.1 is not the only method of determining if a sequence of numbers is graphical. The next result gives another characterization of graphical sequences (Erdős and Gallai, 1960).

Theorem 2.3.2. (Erdös–Gallai Theorem) *Let* $d_1 \geq d_2 \geq \cdots \geq d_n$ *be a sequence of non-negative integers, where* $d_1 \geq 1$ *and* $n \geq 2$. *Then* $d_1, d_2, \ldots, d_n$ *is a graphical sequence if and only if*

(i) $\displaystyle\sum_{i=1}^{n} d_i$ *is even; and*

(ii) $\displaystyle\sum_{i=1}^{k} d_i \leq k(k-1) + \sum_{i=k+1}^{n} \min\{d_i, k\}$ *for* $k = 1, \ldots, n$.

One direction is both clever and easy. Suppose G is a graph on n vertices with degree sequence $d_1, d_2, \ldots, d_n$. Observe that (i) follows from Proposition 1.1.1. To prove (ii) choose $k \in \{1, 2, \ldots, n\}$ and partition the vertex set V into two disjoint sets $\{v_1, \ldots, v_k\}$ and $\{v_{k+1}, \ldots, v_n\}$. Consider the inequality

$$\sum_{i=1}^{k} d_i \leq k(k-1) + \sum_{i=k+1}^{n} \min\{d_i, k\}.$$

The sum on the left is the sum of the degrees of $v_1, \ldots, v_k$. There can be no more than $k(k-1)$ edges connecting $v_1, \ldots, v_k$ to each other. Moreover, a vertex v_i, where $i > k$ cannot connect to more than k vertices in $\{v_1, \ldots, v_k\}$, and it also obviously cannot connect to more than d_i vertices. So its contribution is no more than $min\{d_i, k\}$. Therefore, there is a contribution of at most $k(k-1)$ to this sum from edges in $\{v_1, \ldots v_k\}$ and a contribution of at most $min\{k, d_i\}$ from vertices in $\{v_{k+1}, \ldots, v_n\}$. The other direction is more complicated. A proof may be found in Tripathi et al. (2010).

The next result gives a necessary and sufficient condition for a sequence of ordered pairs to be the degree sequence of a digraph.[5]

Theorem 2.3.3. *A sequence of ordered pairs of non-negative integers* (r_1, s_1), $(r_2, s_2), \ldots, (r_n, s_n)$, *where* $r_1 \geq r_2 \geq \cdots \geq r_n$ *is the degree sequence of a digraph if and only if*

(i) $\displaystyle\sum_{i=1}^{n} r_i = \sum_{i=1}^{n} s_i;$ *and*

(ii) $\displaystyle\sum_{i=1}^{k} r_i \leq \sum_{i=1}^{k} min\{s_i, k-1\} + \sum_{i=k+1}^{n} min\{s_i, k\}$ *for* $1 \leq k \leq n.$

Again one direction is easy as compared to the converse. Suppose D is a digraph on n vertices with degree sequence $(r_1, s_1), (r_2, s_2), \ldots, (r_n, s_n)$. Observe that (i) follows from Proposition 1.1.3. To prove (ii) choose $k \in \{1, 2, \ldots, n\}$ and partition the vertex set V into two disjoint sets $\{v_1, \ldots, v_k\}$ and $\{v_{k+1}, \ldots v_n\}$. Consider the inequality

$$\sum_{i=1}^{k} r_i \leq \sum_{i=1}^{k} min\{s_i, k-1\} + \sum_{i=k+1}^{n} min\{s_i, k\}.$$

The sum on the left is the sum of the indegrees of arcs with initial vertex in $\{v_1, \ldots, v_k\}$. A vertex $v_i \in \{v_1, \ldots, v_k\}$ serves as the terminal vertex of at most $min\{s_i, k-1\}$ of these arcs. A vertex $v_i \in \{v_{k+1}, \ldots, v_n\}$ serves as the terminal vertex of at most $min\{s_i, k\}$ of these arcs. Therefore, there is a contribution of at most $\displaystyle\sum_{i=1}^{k} min\{s_i, k-1\}$ to this sum from vertices in $\{v_1, \ldots, v_k\}$ and a contribution of at most $\displaystyle\sum_{i=k+1}^{n} min\{s_i, k\}$ from vertices in $\{v_{k+1}, \ldots, v_n\}$.

---

5 The manner in which the final form of the theorem emerged is an interesting example of how incremental improvements are made to theorems. The first person to prove the result was Fulkerson (1960). Later Chen (1966) showed that the ordered pairs could be arranged in decreasing lexicographic order, thereby reducing the number of inequalities that must be considered. Then Anstee (1982) showed that it was enough to state $r_1 \geq \cdots \geq r_n$ in the hypothesis. See Berger (2014) for the history and a proof.

2.4 Matrices

In this section we will study three different types of matrices associated with a graph: adjacency matrices, incidence matrices, and Laplacian matrices. This section requires a good understanding of linear algebra. See Appendix A for a brief introduction and an explanation of the terminology used.

The most commonly used matrix to represent a graph is the adjacency matrix. Let G be a graph with n vertices $v_1, \ldots, v_n$. The *adjacency matrix* of G is an $n \times n$ matrix $A = [a_{ij}]$ with vertices along the rows and columns. The ijth entry $a_{ij} = 1$ if $v_i v_j$ is an edge and 0 otherwise. The *adjacency matrix* of a digraph with n vertices is defined in a similar manner except that edges are replaced with directed edges (arcs). The ijth entry $a_{ij} = 1$ if (v_i, v_j) is an arc and 0 otherwise. For example, the adjacency matrices of the graph G and digraph D in Figure 1.3, drawn again for convenience, are shown in the following text.

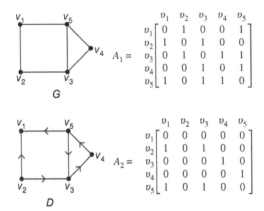

$$A_1 = \begin{array}{c} \\ v_1 \\ v_2 \\ v_3 \\ v_4 \\ v_5 \end{array} \begin{array}{ccccc} v_1 & v_2 & v_3 & v_4 & v_5 \\ \begin{bmatrix} 0 & 1 & 0 & 0 & 1 \\ 1 & 0 & 1 & 0 & 0 \\ 0 & 1 & 0 & 1 & 1 \\ 0 & 0 & 1 & 0 & 1 \\ 1 & 0 & 1 & 1 & 0 \end{bmatrix} \end{array}$$

$$A_2 = \begin{array}{c} \\ v_1 \\ v_2 \\ v_3 \\ v_4 \\ v_5 \end{array} \begin{array}{ccccc} v_1 & v_2 & v_3 & v_4 & v_5 \\ \begin{bmatrix} 0 & 0 & 0 & 0 & 0 \\ 1 & 0 & 1 & 0 & 0 \\ 0 & 0 & 0 & 1 & 0 \\ 0 & 0 & 0 & 0 & 1 \\ 1 & 0 & 1 & 0 & 0 \end{bmatrix} \end{array}$$

The adjacency matrix of a graph is symmetric with zeros along the diagonal. The adjacency matrix of a digraph has zeros along the diagonal, but it is not symmetric. In the matrix A_2 corresponding to digraph D in Figure 1.3, there is no arc pointing away from vertex v_1; hence the row of zeros. There are two arcs pointing away from v_2, namely (v_2, v_1) and (v_2, v_3); hence the two ones in the second row.

The adjacency matrix encapsulates valuable information about the graph, some of it obvious and some hidden. There are three immediate observations we can make.

(1) For a graph, $deg(v_i)$ is the sum of the ones in row v_i (or column v_i) of the adjacency matrix. We write

$$deg(v_i) = \sum_{j=1}^{n} a_{ij} = \sum_{j=1}^{n} a_{ji}.$$

For a digraph, $outdeg(v_i)$ is the sum of the ones in row v_i and $indeg(v_i)$ is the sum of the ones in column v_i. We write

$$outdeg(v_i) = \sum_{j=1}^{n} a_{ij}$$

and

$$indeg(v_i) = \sum_{j=1}^{n} a_{ji}.$$

(2) The degree sequence may be viewed as a vector $\bar{x}_{deg}$ consisting of the vertex degrees. We write

$$\bar{x}_{deg} = \begin{bmatrix} deg(v_1) \\ deg(v_2) \\ \vdots \\ deg(v_n) \end{bmatrix}.$$

Denote by $\bar{1}_n$ the vector of all ones (read as "one bar"). Then the degree sequence may be obtained as $A\bar{1}_n = \bar{x}_{deg}$. For example, the degree sequence for the graph in Figure 1.3, redrawn with degrees by each vertex, may be obtained as follows:

$$\begin{bmatrix} 0 & 1 & 0 & 0 & 1 \\ 1 & 0 & 1 & 0 & 0 \\ 0 & 1 & 0 & 1 & 1 \\ 0 & 0 & 1 & 0 & 1 \\ 1 & 0 & 1 & 1 & 0 \end{bmatrix} \begin{bmatrix} 1 \\ 1 \\ 1 \\ 1 \\ 1 \end{bmatrix} = \begin{bmatrix} 2 \\ 2 \\ 3 \\ 2 \\ 3 \end{bmatrix}.$$

Similarly for digraphs, if A is the adjacency matrix of a digraph, then

$$A\bar{1}_n = \bar{x}_{outdeg}$$

and

$$A^T\bar{1}_n = \bar{x}_{indeg}$$

The outdegree sequence for the digraph in Figure 1.3, redrawn with pairs of numbers indicating indegree and outdegree by each vertex may be obtained as follows:

$$\begin{bmatrix} 0 & 0 & 0 & 0 & 0 \\ 1 & 0 & 1 & 0 & 0 \\ 0 & 0 & 0 & 1 & 0 \\ 0 & 0 & 0 & 0 & 1 \\ 1 & 0 & 1 & 0 & 0 \end{bmatrix} \begin{bmatrix} 1 \\ 1 \\ 1 \\ 1 \\ 1 \end{bmatrix} = \begin{bmatrix} 0 \\ 2 \\ 1 \\ 1 \\ 2 \end{bmatrix}.$$

The indegree sequence may be obtained as follows:

$$
\begin{bmatrix}
0 & 1 & 0 & 0 & 1 \\
0 & 0 & 0 & 0 & 0 \\
0 & 1 & 0 & 0 & 1 \\
0 & 0 & 1 & 0 & 0 \\
0 & 0 & 0 & 1 & 0
\end{bmatrix}
\begin{bmatrix}
1 \\ 1 \\ 1 \\ 1 \\ 1
\end{bmatrix}
=
\begin{bmatrix}
2 \\ 0 \\ 2 \\ 1 \\ 1
\end{bmatrix}.
$$

(3) The diagonal of the square of the adjacency matrix gives the degree sequence of the graph. Let b_{ij} be the ijth entry of A^2. Since A is symmetric, and consequently $a_{ik} = a_{ki}$ is one or zero,

$$
b_{ii} = \sum_{k=1}^{n} a_{ik} a_{ki} = \sum_{k=1}^{n} a_{ik} = deg(v_i).
$$

Thus when multiplying A by itself, entries along the diagonal of A^2 form the degree sequence. For example,

$$
A_1^2 =
\begin{array}{c}
\\ v_1 \\ v_2 \\ v_3 \\ v_4 \\ v_5
\end{array}
\begin{array}{c}
\begin{array}{ccccc}
v_1 & v_2 & v_3 & v_4 & v_5
\end{array} \\
\begin{bmatrix}
2 & 0 & 2 & 1 & 0 \\
0 & 2 & 0 & 1 & 2 \\
2 & 0 & 3 & 1 & 1 \\
1 & 1 & 1 & 2 & 1 \\
0 & 2 & 1 & 1 & 3
\end{bmatrix}
\end{array},
$$

and the degree sequence 2, 2, 3, 2, 3 appears in the diagonal. Matrix A_1^2 captures additional useful information about the graph. There are two $v_1 - v_1$ walks of length 2, zero $v_1 - v_2$ walks of length 2, two $v_1 - v_3$ walks of length 2, and so on. These are not coincidences as the next proposition establishes.

Proposition 2.4.1. *Let G be a graph with vertices $v_1, \ldots, v_n$ and let A be its adjacency matrix. The ijth entry of A^k, where $k \geq 1$, is the number of distinct $v_i - v_j$ walks of length k.*

Proof. The proof is by induction on k. If $k = 1$, by definition the ijth entry of A is 1 if there is an edge between v_i and v_j, and 0 otherwise. So the ijth entry may be viewed as the number of $v_i - v_j$ walks of length 1. Suppose the result is true for $k - 1$. That is, the ijth entry of A^{k-1} (call it b_{ij}) is the number of $v_i - v_j$ walks of length $k - 1$. Since $A^k = A^{k-1}A$, the ijth entry of A^k is

$$
\sum_{t=1}^{n} b_{it} a_{tj} = b_{i1} a_{1j} + \cdots + b_{in} a_{nj},
$$

where b_{it} represents the number of $v_i - v_t$ walks of length $k - 1$ and a_{tj} represents edge $v_t v_j$. Observe that $a_{tj} = 1$ if edge $v_t v_j$ exists and 0 otherwise. Therefore $\sum_{t=1}^{n} b_{it} a_{tj}$ represents the number of $v_i - v_j$ walks of length k. $\qquad\square$

Proposition 2.4.1 can be used to find the number of triangles in a graph. The sum of the entries of the diagonal of a square matrix is called the *trace* of the matrix and is denoted by $tr(A)$.

Corollary 2.4.2. *Let G be a graph with adjacency matrix A. The number of triangles in G is $\frac{1}{6}tr(A^3)$.*

Proof. Suppose G is a graph. A triangle $\{v_i, v_j, v_k, v_i\}$ may be viewed as a $v_i - v_i$ walk of length 3. By Proposition 2.4.1 the number of $v_i - v_i$ walks of length 3 is the sum of the entries in the diagonal of A^3. Therefore, the number of triangles is $tr(A^3)$. However, each $v_i - v_i$ walk is counted three times for each of the three vertices in the triangle, and in both directions. So, we divide by $3 \times 2 = 6$. Therefore the number of triangles is $\frac{1}{6}tr(A^3)$. $\qquad\square$

The conclusions in Proposition 2.4.1 and Corollary 2.4.2 also hold for digraphs. The arguments are similar to those in the previous proofs.

The problem of finding directed cycles in digraphs leads to an interesting unresolved conjecture (Caccetta and Häggkvist, 1978).

Conjecture 2.4.3. (Caccetta-Häggkvist Conjecture) If G is a digraph on n vertices with minimum out-degree s, then G has a directed cycle of length at most $\lceil \frac{n}{s} \rceil$.

Let G be a graph on n vertices and let A be its adjacency matrix. An *eigenvector* of A is a non-zero vector $\bar{x}$ such that $A\bar{x} = \lambda\bar{x}$, where λ is a real number. We call λ the *eigenvalue* corresponding to *eigenvector* $\bar{x}$. The eigenvalues of a matrix are called its *spectrum*. Spectral graph theory is the study of the relationships between a graph's structure and the eigenvalues of a matrix associated with it.

The next proposition establishes that isomorphic graphs have the same eigenvalues. This is not an obvious statement, especially since the adjacency matrix depends on the labels of the vertices. We begin by illustrating the ideas used in the proof with an example. Consider the two isomorphic graphs G_1 and G_2 shown in Figure 1.15 with adjacency matrices

$$
A = \begin{array}{c} \\ 1 \\ 2 \\ 3 \\ 4 \end{array}
\begin{array}{cccc} 1 & 2 & 3 & 4 \\ \left[\begin{array}{cccc} 0 & 1 & 1 & 1 \\ 1 & 0 & 1 & 0 \\ 1 & 1 & 0 & 1 \\ 1 & 0 & 1 & 0 \end{array}\right] \end{array}
\qquad
B = \begin{array}{c} \\ 1 \\ 2 \\ 3 \\ 4 \end{array}
\begin{array}{cccc} 1 & 2 & 3 & 4 \\ \left[\begin{array}{cccc} 0 & 1 & 0 & 1 \\ 1 & 0 & 1 & 1 \\ 0 & 1 & 0 & 1 \\ 1 & 1 & 1 & 0 \end{array}\right] \end{array}.
$$

The isomorphism from G_1 to G_2 is given by the permutation.

$$\begin{pmatrix} 1 & 2 & 3 & 4 \\ 2 & 1 & 4 & 3 \end{pmatrix}.$$

Switching rows in a matrix corresponds to multiplying the matrix on the left by a permutation matrix.[6] Switching columns corresponds to multiplying on the right by a permutation matrix. The permutation matrix in this case is

$$P = \begin{bmatrix} 0 & 1 & 0 & 0 \\ 1 & 0 & 0 & 0 \\ 0 & 0 & 0 & 1 \\ 0 & 0 & 1 & 0 \end{bmatrix}.$$

Multiplying A by P on the left results in switching rows 1 and 2 and rows 3 and 4.

$$PA = \begin{bmatrix} 0 & 1 & 0 & 0 \\ 1 & 0 & 0 & 0 \\ 0 & 0 & 0 & 1 \\ 0 & 0 & 1 & 0 \end{bmatrix} \begin{bmatrix} 0 & 1 & 1 & 1 \\ 1 & 0 & 1 & 0 \\ 1 & 1 & 0 & 1 \\ 1 & 0 & 1 & 0 \end{bmatrix} = \begin{bmatrix} 1 & 0 & 1 & 0 \\ 0 & 1 & 1 & 1 \\ 1 & 0 & 1 & 0 \\ 1 & 1 & 0 & 1 \end{bmatrix}.$$

Multiplying PA by P^T on the right results in columns being switched.

$$PAP^T = \begin{bmatrix} 1 & 0 & 1 & 0 \\ 0 & 1 & 1 & 1 \\ 1 & 0 & 1 & 0 \\ 1 & 1 & 0 & 1 \end{bmatrix} \begin{bmatrix} 0 & 1 & 1 & 1 \\ 1 & 0 & 1 & 0 \\ 1 & 1 & 0 & 1 \\ 1 & 0 & 1 & 0 \end{bmatrix} = \begin{bmatrix} 0 & 1 & 0 & 1 \\ 1 & 0 & 1 & 1 \\ 0 & 1 & 0 & 1 \\ 1 & 1 & 1 & 0 \end{bmatrix} = B.$$

Thus the effect of multiplying A by P on the left and P^T on the right is to permute the rows and columns of A in order to obtain B.

Proposition 2.4.4. *The adjacency matrices of isomorphic graphs have the same eigenvalues.*

Proof. Let G and H be isomorphic graphs with adjacency matrices A and B, respectively. We will begin by showing that $B = PAP^T$, where P is a permutation matrix. Since G and H are isomorphic graphs there is a permutation ψ on the set of vertices such that $A = [a_{ij}]$ if and only if $B = [a_{\psi(i)\psi(j)}]$. Construct a permutation matrix $P = [p_{ij}]$, where $p_{ij} = 1$ if $\psi(i) = j$ and zero otherwise. Observe that the

---

6 A permutation matrix P is a matrix that is obtained from an identity matrix with rows switched. It is easy to check that $P^{-1} = P^T$. Two matrices A and B are similar if there exists an invertible matrix C such that $B = CAC^{-1}$. Similar matrices have the same eigenvalues (Proposition A.2).

ijth entry of PA is $\sum_{k=1}^{n} p_{ik}a_{kj} = a_{\psi(i)j}$ since $p_{ik} = 1$ if $\psi(i) = k$ and zero otherwise. Similarly, the ijth entry of $(PA)P^T$ is

$$\sum_{k=1}^{n} a_{\psi(i)k}p_{jk} = a_{\psi(i)\psi(j)}$$

since the kjth entry of P^T is p_{jk} and

$$p_{jk} = \begin{cases} 1 & \text{if } \psi(j) = k \\ 0 & \text{otherwise} \end{cases}.$$

Therefore $B = PAP^T$. Since P is a permutation matrix, $P^T = P^{-1}$ and we may write $B = PAP^{-1}$. Therefore matrices A and B are similar. Proposition A.2 implies that similar matrices have the same eigenvalues. □

Eigenvalues do not uniquely determine the graph. For example, consider the star graph $K_{1,4}$ and the cycle of length 4 with an isolated vertex in the middle shown in Figure 2.7. The adjacency matrices for these two graphs X and Y are shown below:

$$X = \begin{array}{c} \\ v_1 \\ v_2 \\ v_3 \\ v_4 \\ v_5 \end{array} \begin{array}{c} \begin{array}{ccccc} v_1 & v_2 & v_3 & v_4 & v_5 \end{array} \\ \begin{bmatrix} 0 & 1 & 1 & 1 & 1 \\ 1 & 0 & 0 & 0 & 0 \\ 1 & 0 & 0 & 0 & 0 \\ 1 & 0 & 0 & 0 & 0 \\ 1 & 0 & 0 & 0 & 0 \end{bmatrix} \end{array} \quad Y = \begin{array}{c} \\ v_1 \\ v_2 \\ v_3 \\ v_4 \\ v_5 \end{array} \begin{array}{c} \begin{array}{ccccc} v_1 & v_2 & v_3 & v_4 & v_5 \end{array} \\ \begin{bmatrix} 0 & 0 & 0 & 0 & 0 \\ 0 & 0 & 1 & 0 & 1 \\ 0 & 1 & 0 & 1 & 0 \\ 0 & 0 & 1 & 0 & 1 \\ 0 & 1 & 0 & 1 & 0 \end{bmatrix} \end{array}.$$

The eigenvalues for both matrices X and Y are 2, −2, and 0 (repeated three times). Thus these two graphs have the same spectrum, but they are clearly not isomorphic. Non-isomorphic graphs with the same eigenvalues are called *cospectral* (with respect to the matrix under consideration).

The next result lists several properties of the spectrum. They depend on linear algebra results in Appendix A.

Figure 2.7 Cospectral graphs with respect to the adjacency matrix.

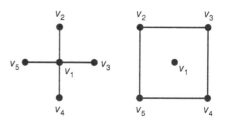

Proposition 2.4.5. *Let G be a graph with n vertices, m edges, and adjacency matrix A with eigenvalues $\lambda_1, \ldots, \lambda_n$. Then*

(i) *$\lambda_1, \ldots, \lambda_n$ are real numbers;*
(ii) *$\lambda_1 + \lambda_2 + \cdots + \lambda_n = 0$; and*
(iii) *$\lambda_1^2 + \lambda_2^2 + \cdots + \lambda_n^2 = 2m$.*

Proof. Since A is a real symmetric matrix, (i) follows immediately from A.4(i). The sum of the eigenvalues of A equals $tr(A)$ by Proposition A.4(ii), which is zero since A is an adjacency matrix. Therefore

$$\lambda_1 + \lambda_2 + \cdots + \lambda_n = 0.$$

Finally, since A is a square matrix, Proposition A.1(ii) implies that the eigenvalues of A^2 are $\lambda_1^2, \ldots, \lambda_n^2$. The diagonal entries of A^2 are the degrees of vertices, so Proposition 1.1.1 implies that

$$\lambda_1^2 + \cdots + \lambda_n^2 = tr(A^2) = \sum deg(v) = 2m. \qquad \square$$

The *incidence matrix B* of a graph (or digraph) has vertices along the rows and edges along the columns. For a graph, entries b_{ij} are defined as

$$b_{ij} = \begin{cases} 1 & \text{if } e_j \text{ is incident to } v_i \\ 0 & \text{otherwise} \end{cases}$$

For a digraph, entries b_{ij} are defined as

$$b_{ij} = \begin{cases} 1 & \text{if } e_j \text{ enters } v_i \\ -1 & \text{if } e_j \text{ leaves } v_i \\ 0 & \text{otherwise} \end{cases}.$$

For example, the incidence matrices of the graph G and digraph D in Figure 1.3, drawn again for convenience, are shown in the following text.

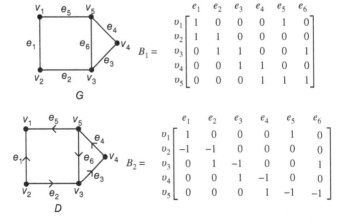

Let G be a graph on n vertices with adjacency matrix A and let $D = [d_{ij}]$ be a diagonal matrix with diagonal entries $d_{ii} = deg(v_i)$. The *Laplacian matrix* is defined as $L = D - A$. See Mohar et al. (1991) for a survey on the Laplacian matrix and its applications. For example, the Laplacian matrix of graph G in Figure 1.3 is obtained as follows:

$$L = \begin{bmatrix} 2 & 0 & 0 & 0 & 0 \\ 0 & 2 & 0 & 0 & 0 \\ 0 & 0 & 3 & 0 & 0 \\ 0 & 0 & 0 & 2 & 0 \\ 0 & 0 & 0 & 0 & 3 \end{bmatrix} - \begin{bmatrix} 0 & 1 & 0 & 0 & 1 \\ 1 & 0 & 1 & 0 & 0 \\ 0 & 1 & 0 & 1 & 1 \\ 0 & 0 & 1 & 0 & 1 \\ 1 & 0 & 1 & 1 & 0 \end{bmatrix} = \begin{bmatrix} 2 & -1 & 0 & 0 & -1 \\ -1 & 2 & -1 & 0 & 0 \\ 0 & -1 & 3 & -1 & -1 \\ 0 & 0 & -1 & 2 & -1 \\ -1 & 0 & -1 & -1 & 3 \end{bmatrix}.$$

Observe that entries l_{ij} in L are given by

$$l_{ij} = \begin{cases} deg(v_i) & \text{if } i = j \\ -1 & \text{if } i \neq j \text{ and } (v_i, v_j) \text{ is an edge} \\ 0 & \text{otherwise} \end{cases}.$$

As with the adjacency matrix, there are relationships between the structure of a graph and the eigenvalues of the Laplacian matrix. The Laplacian matrix is not only a real symmetric matrix like the adjacency matrix, but also positive semi-definite.[7] It is fairly straightforward to check that $L = BB^T$, where B is the incidence matrix of the graph with an arbitrary assignment of arrows on the edges. For example, the digraph in Figure 1.3 may be viewed as the graph next to it with an arbitrary assignment of arrows on the edges. The digraph has incidence matrix B_2 listed earlier. Observe that

$$L = \begin{bmatrix} 2 & -1 & 0 & 0 & -1 \\ -1 & 2 & -1 & 0 & 0 \\ 0 & -1 & 3 & -1 & -1 \\ 0 & 0 & -1 & 2 & -1 \\ -1 & 0 & -1 & -1 & 3 \end{bmatrix}$$

$$= \begin{bmatrix} 1 & 0 & 0 & 0 & 1 & 0 \\ -1 & -1 & 0 & 0 & 0 & 0 \\ 0 & 1 & -1 & 0 & 0 & 1 \\ 0 & 0 & 1 & -1 & 0 & 0 \\ 0 & 0 & 0 & 1 & -1 & -1 \end{bmatrix} \begin{bmatrix} 1 & -1 & 0 & 0 & 0 \\ 0 & -1 & 1 & 0 & 0 \\ 0 & 0 & -1 & 1 & 0 \\ 0 & 0 & 0 & -1 & 1 \\ 1 & 0 & 0 & 0 & -1 \\ 0 & 0 & 1 & 0 & -1 \end{bmatrix}.$$

---

7 A symmetric matrix with positive eigenvalues is called *positive definite*. A symmetric matrix with non-negative eigenvalues is called *positive semidefinite*. Equivalently, a symmetric matrix M is positive semidefinite if it can be written as $M = UU^T$ for some matrix U.

Since L is positive semidefinite, the eigenvalues of L are non-negative real numbers. The next proposition lists additional properties of the Laplacian matrix.

Proposition 2.4.6. *Let G be a graph with n vertices and m edges, and let $\lambda_1, \lambda_2, \ldots, \lambda_n$ be the eigenvalues of the Laplacian matrix L. Then*

(i) $\lambda_1 + \cdots + \lambda_n = 2m$;
(ii) 0 is an eigenvalue of L with corresponding eigenvector $\overline{1}_n$; and
(iii) The multiplicity of 0 is the number of components in G.

Proof. Observe that (i) follows from Corollary A.4(ii) and the fact that $tr(L) = 2m$. To prove (ii) observe that the sum of the entries in each row is 0. Therefore

$$L\overline{1}_n = \overline{0} = 0\overline{1}_n.$$

Consequently, 0 is an eigenvalue corresponding to eigenvector $\overline{1}_n$. To prove (iii) suppose G is disconnected with t connected components and $L_1, \ldots, L_t$ are the Laplacian matrices of the t components. Then L consists of $L_1, \ldots L_t$ arranged along the diagonal with zeros elsewhere. Therefore L is a block diagonal matrix.[8] Each L_i contributes one zero to the eigenvalues of L by Part (ii). So 0 is an eigenvalue of L with multiplicity t. □

Let G be a graph with n vertices and Laplacian matrix L. Since the eigenvalues of L are non-negative, we can write them in order as

$$\lambda_1 \le \lambda_2 \le \cdots \le \lambda_n.$$

If G is a connected graph, then $\lambda_1 = 0$ and $\lambda_2 \ne 0$ by Proposition 2.4.6(ii) and (iii). The second smallest eigenvalue, λ_2, is called the *algebraic connectivity* (Fiedler, 1973) and an eigenvector corresponding to it is called a *Fiedler vector*. The algebraic connectivity is a more granular measure of connectivity and depends on the vertices as well as the manner in which vertices are linked.

For example, the eigenvalues of the Laplacian matrix of the graph G in Figure 1.3 are

$$\lambda_1 = 0, \quad \lambda_2 \approx 1.38, \quad \lambda_3 \approx 2.38, \quad \lambda_4 \approx 3.62, \quad \lambda_5 \approx 4.62.$$

The algebraic connectivity of this graph is 1.38.

We will end this section with the well-known Matrix Tree Theorem. The first minor M_{ij} of a matrix is obtained by deleting the ith row and jth column. Kirchhoff (1847) found a remarkable result relating the eigenvalues of the Laplacian matrix

---

8 A square matrix B in which each diagonal entry is itself a square matrix and each off-diagonal entry is 0 is called a *block diagonal matrix*. It can be shown that the eigenvalues of the square matrices along the diagonal are the eigenvalues of B.

with the number of spanning trees in the graph. A proof of this result may be found in Rényi (1970) or Chaiken and Kleitman (1978).

Theorem 2.4.7. (Kirchhoff's Matrix Tree Theorem) *Let G be a graph and let M_{ij} be a first minor of the Laplacian matrix of G. The number of spanning trees in G is $|det(M_{ij})|$.*

For example, as noted earlier the Laplacian matrix of the graph G in Figure 1.3 is

$$
L = \begin{bmatrix}
2 & -1 & 0 & 0 & -1 \\
-1 & 2 & -1 & 0 & 0 \\
0 & -1 & 3 & -1 & -1 \\
0 & 0 & -1 & 2 & -1 \\
-1 & 0 & -1 & -1 & 3
\end{bmatrix}.
$$

Any first minor of L may be considered (i.e. any row and column may be deleted.) The first minor $M_{2,1}$ shown below is obtained from L by deleting the second row and first column.

$$
M_{2,1} = \begin{bmatrix}
-1 & 0 & 0 & -1 \\
-1 & 3 & -1 & -1 \\
0 & -1 & 2 & -1 \\
0 & -1 & -1 & 3
\end{bmatrix}.
$$

Theorem 2.4.7 implies that the number of spanning trees is $|det(M_{2,1})| = 11$. This is a small graph and the 11 spanning trees may be found easily. They are:

$$\{e_1,e_2,e_3,e_4\}\{e_1,e_2,e_3,e_5\}\{e_1,e_2,e_3,e_6\}\{e_1,e_2,e_4,e_5\}\{e_1,e_2,e_4,e_6\}\{e_1,e_3,e_4,e_5\}$$

$$\{e_1,e_3,e_5,e_6\}\{e_1,e_4,e_5,e_6\}\{e_2,e_3,e_4,e_5\}\{e_2,e_3,e_5,e_6\}\{e_2,e_4,e_5,e_6\}.$$

These are just a small selection of spectral graph theory results presented for their intrinsic value and because they are used in Sections 3.3 and 3.4.

Exercises

2.1 For each of the graphs in Figure 2.8:
(a) Find the eccentricity and total distance of each vertex, and the radius and diameter of the graph.
(b) Find the center, periphery, and median.
(c) Find the mean distance.
(d) Find the adjacency, incidence, and Laplacian matrices.
(e) Find the number of $v_i - v_j$ walks of lengths 2, 3, and 4.

(f) Find the eigenvalues of the adjacency matrix.
(g) Find the eigenvalues of the Laplacian matrix and the algebraic connectivity.
(h) Use Theorem 2.4.7 (Matrix Tree Theorem) to find the number of spanning trees.

2.2 For the strongly connected digraphs in Figure 2.8 define $d(u, v)$ as the shortest directed path from u to v.
(a) Find the eccentricity and total distance of each vertex.
(b) Find the mean distance.
(c) Find the adjacency matrix.
(d) Find the number of $v_i - v_j$ directed walks of lengths 2, 3, and 4.

2.3 Use Theorem 2.3.1 to determine if the following sequences of numbers are graphical. If they are graphical, then construct a graph with the degree sequence.
(a) $3, 3, 2, 2$
(b) $5, 5, 3, 3, 2, 2, 2$
(c) $5, 5, 5, 3, 3, 3, 3, 3$
(d) $5, 4, 3, 1, 1, 1, 1, 1, 1$
(e) $5, 5, 3, 3, 2, 2, 2$

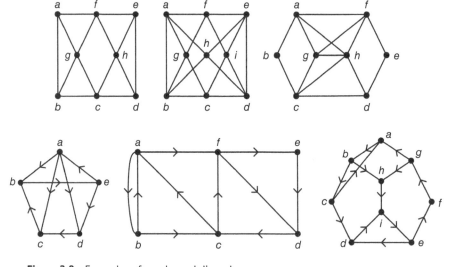

Figure 2.8 Examples of graphs and digraphs.

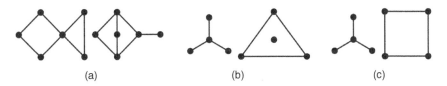

Figure 2.9 Three pairs of cospectral graphs.

2.4 We say a graph is *minimally connected* if it is connected, but removal of any edge disconnects it. Prove that a graph is a tree if and only if it is minimally connected.

2.5 Prove that the center of a tree is a vertex or an edge (Jordan, 1869).

2.6 Prove that if G is a tree, then $diam(G)$ is either $2rad(G)$ or $2rad(G) - 1$.

2.7 Prove that if G is a simple graph with $diam(G) \geq 3$, then $diam(\overline{G}) \leq 3$.

2.8 Prove that a tree is reconstructable from its deck of vertex-deletions (Kelly, 1957).

2.9 Let G be a connected graph and v be a vertex that is not a cut vertex and e be an edge that is not a bridge. Prove the following statements.
(a) $rad(G) - 1 \leq rad(G - v)$.
(b) $diam(G) - 1 \leq diam(G - v)$.
(c) $rad(G) \leq rad(G \backslash e) \leq 2rad(G)$.
(d) $diam(G) \leq diam(G \backslash e) \leq 2diam(G)$.

2.10 The *signless Laplacian matrix* is defined as $Q = D + A$, where D is the diagonal matrix with vertex degrees along the diagonal and A is the adjacency matrix. The *normalized Laplacian matrix* is defined as $N = D^{-\frac{1}{2}}(D - A)D^{-\frac{1}{2}}$. Find Q and N for the graphs in Figure 2.8. Verify that the graphs (a), (b), and (c) in Figure 2.9 are cospectral for the Laplacian matrix, the signless Laplacian, and the normalized Laplacian, respectively.

Topics for Deeper Study

2.11 Review the many proofs for Theorem 2.1.7 (Cayley's Tree Counting Theorem). Four different proofs are presented in (Aigner and Ziegler, 2010) and additional proofs are in *Counting Labeled Trees* (Moon, 1970).

2.12 Let G be a strongly connected digraph. The directed distance between a pair of vertices u and v, defined as the length of the shortest directed path, is not a metric (Why?). In a paper titled "Distance in Digraphs," the authors defined two new types of directed distances that are metrics. In addition, they summarized several previous results on the center $C(G)$, median $M(G)$, and periphery $P(G)$ of a graph (Chartrand and Tian, 1997). Prove the following statements.

(a) The center of a connected graph G lies in a single block (Harary and Norman, 1953).

(b) For every graph G, there is a connected graph H with $C(H) \cong G$ (Buckley et al., 1981).

(c) For every graph G, there is a connected graph H with $M(H) \cong G$ (Slater, 1980).

(d) For every graph G, there is a connected graph H such that $P(H) \cong G$ if and only if every vertex of G has eccentricity 1 or no vertex has eccentricity 1 (Bielak and Syslo, 1983).

(e) For every pair of graphs G_1 and G_2, there is a connected graph H such that $C(H) \cong G_1$ and $M(H) \cong G_2$ (Hendry, 1985).

(f) Let H_1 and H_2 be two subgraphs of a connected graph. Define the distance between H_1 and H_2 as

$$d(H_1, H_2) = min\{d(v_1, v_2) \mid v_1 \in V(H_1), v_2 \in V(H_2)\}.$$

For every pair of graphs G_1 and G_2 and positive integer k, there exists a connected graph H with $C(H) \cong G_1$, $M(H) \cong G_2$, and $d(C(H), M(H)) = k$ (Holbert, 1989).

(g) For every pair graphs G_1 and G_2 and every graph K that is isomorphic to an induced subgraph of G_1 and G_2, there is a connected graph H with $C(H) \cong G_1$, $M(H) \cong G_2$, and $C(H) \cap M(H) \cong K$ (Novotny and Tian, 1991).

2.13 In a survey paper titled "Distance in graphs" the authors gather together several results related to distance (Goddard and Oellermann, 2011). Let G be a connected graph with n vertices and maximum degree Δ. Prove the following statements:

(a) $diam(G) \leq n - \Delta + 1$.

(b) $rad(G) \leq \frac{n-\Delta}{2} + 1$.

(c) If $\delta(G) \geq \frac{n}{2}$, then $diam(G) \leq 2$.

(d) Give examples to show that these bounds are sharp.

2.14 Let G be a connected graph with n vertices and minimum degree $\delta \geq 2$. Prove the following statements (Erdős et al., 1989).

(a) $rad(G) \leq \frac{3n-3}{2\delta+1} + 5.$

(b) If G has no triangles, then
- $diam(G) \leq 4 \left\lfloor \frac{n-\delta-1}{2\delta} \right\rfloor$; and
- $rad(G) \leq \frac{n-2}{\delta} + 12.$

(c) If G has no cycles of length 4, then
- $diam(G) \leq \frac{5n}{\delta^2 - 2\lfloor \frac{\delta}{2} \rfloor + 1}$; and
- $rad(G) \leq \frac{5n}{2(\delta^2 - 2\lfloor \frac{\delta}{2} \rfloor + 1)}.$

3

Similarity and Centrality

Measures that highlight the similarity between two vertices are called *vertex similarity measures*. Measures that highlight the differences between vertices are called *vertex centrality measures*. Section 3.1, is on similarity measures for vertices and Sections 3.2 – 3.4 are on centrality measures. In Section 2.2 we defined important vertices in two ways: central vertices and median vertices. The approach in this chapter is quite different. In Section 3.2 we will describe how to assign a number to each vertex indicating its importance. These measures are based on paths between pairs of vertices. In Section 3.3 we will do the same, but the measures will be based on walks between pairs of vertices. Section 3.4 is on PageRank, the webpage ranking method invented by the founders of Google.

3.1 Similarity Measures

Recall from Section 1.2 that an automorphism of a graph is an isomorphism between a graph and itself. Vertices that are mapped to each other by an automorphism of the graph are the same from a structural perspective. However, using automorphism as a similarity measure is not practical when the graphs are large, incomplete, or frequently changing. Moreover, an automorphism does not assign to each pair of vertices a number that can be used to compare pairs of vertices. So, alternate definitions of similarity are needed.

Let G be a connected graph and let u and v be two vertices of G. The *neighborhood similarity* [1] of u and v, denoted by $s(u, v)$, is

---

1 French botanist Paul Jaccard introduced a simple way of comparing two sets (Jaccard, 1901). Suppose A and B are two sets. The similarity between A and B is defined as $\frac{|A \cap B|}{|A \cup B|}$. For example, if sets A and B represent documents and the words in A and B are the elements of the sets, then this quantity is a measure of how many common words are in the documents even though they may be on different topics.

Graphs and Networks, First Edition. S. R. Kingan.
© 2022 John Wiley & Sons, Inc. Published 2022 by John Wiley & Sons, Inc.

$$s(u, v) = \frac{|N(u) \cap N(v)|}{|N(u) \cup N(v)|}.$$

Every pair of vertices u, v has a number $s(u, v) \leq 1$ associated with it, and pairs with values of $s(u, v)$ closer to 1 are more similar to each other.[2]

In 1974 sociologist Mark Granovetter published a monograph titled *Getting A Job: A Study of Contacts and Careers* (Granovetter, 1974). He conducted a study of how 282 men found their jobs, and his analysis supported the hypothesis that people found jobs with the help of a distant acquaintance or a friend of a friend rather than a close friend. He followed up with a paper titled "The strength of weak ties," where he described the importance of having weak friendships with many people from different backgrounds (Granovetter, 1973).[3]

For each edge $e = uv$ in a connected graph, Granovetter assigned the number $s(u, v)$, where $0 \leq s(u, v) \leq 1$. Edges with higher values are considered strong ties whereas edges with lower values are considered weak ties. If $s(u, v) = 0$, which happens when $N(u) \cap N(v) = \phi$, then the edge e is called a *local bridge*.

Local bridges can be distinguished from each other by measuring how much their deletion increases the distance between their end vertices. The *span* of a local bridge $e = uv$, denoted by *span(e)*, is defined as

$$span(e) = d_{G \backslash e}(u, v).$$

Observe that an edge is a local bridge if and only if it is not in any triangle of G. Therefore the span of a local bridge is at least 3. A local bridge with a large span provides its end vertices u and v with access to distant parts of the network.

It had been observed experimentally that over time, if friendship ties linking v_1 to v_2 and v_1 to v_3 are both strong, then a friendship develops between v_2 and v_3 forming a triangle $\{v_1, v_2, v_3\}$. Social scientists call this triadic closure. Social networks with a high prevalence of triadic closure evolve over time to have large cliques. Triadic closure may be good or bad depending on the context. Inner city urban social networks have strong triadic closure and few local bridges; an indication of how hard it is to break free from the cycle of poverty and violence. In this case more local bridges would be better. On the other hand, an analysis of the social network of 13 465 adolescents from the National Longitudinal Survey of Adolescent Health revealed that depression among young women is significantly

---

2 See Leicht et al. (2006) for a history of vertex similarity measures and different ways of defining them. Some authors adopt the convention that when u and v are adjacent to each other, v is not counted in $N(u)$ and u is not counted in $N(v)$.

3 Granovetter and Herbert Gans, author of *Urban Villagers* (Gans, 1962), held an interesting written debate on the strength of weak ties (Gans, 1974). See also Granovetter (1983) and Chapter 3 in Easley and Kleinberg (2010) titled "Strong and Weak Ties."

increased by social isolation and friendship patterns in which friends were not friends with each other. In this case fewer local bridges would be better. No such correlation was found for young men (Bearman and Moody, 2004).

Another approach to comparing vertices is to define a local clustering coefficient for each vertex (Watts and Strogatz, 1998). Let G be a graph and let v be a vertex in G with $deg(v) \geq 2$. The *local clustering coefficient* of v, denoted by $c(v)$, is the number of triangles containing v divided by the number of paths of length 2 with middle vertex v. In the context of social networks, the local clustering coefficient of a vertex (person) is a measure of how likely a person's friends know each other. The number of paths of length 2 with middle vertex v is $\binom{|N(v)|}{2}$. Let H be the subgraph of G induced by $N(v)$. Note that $v \notin N(v)$, so v is not a vertex in H. The number of triangles that contain v is the number of edges that occur in H. Therefore for a vertex v with $deg(v) \geq 2$, the local clustering coefficient is

$$c(v) = \frac{|E(H)|}{\binom{|N(v)|}{2}}.$$

The *mean local clustering coefficient* of G, denoted by $C(G)$, is the mean of the local clustering coefficients, where the mean is taken over vertices of degree at least 2. A large value for the mean local clustering coefficient indicates that the graph is more tightly clustered. This means that if a random person with two friends is selected, then those friends are more likely to know each other. For example, the mean local clustering coefficient of the Erdős-1 collaboration graph in Figure 1.19 is 0.342. Note that the denominator of the clustering coefficient only includes vertices of degree at least 2. So, only coauthors of Erdős who coauthored with at least two other coauthors of Erdős are counted in the denominator.

Alternatively, a "global" clustering coefficient may be defined in a slightly different manner as

$$C_{global}(G) = \frac{3 \times \text{Number of triangles}}{\text{Number of paths of length 3}}.$$

The numerator is multiplied by 3 to balance the denominator, which contains each triangle 3 times. For example, the global clustering coefficient of the Erdős-1 collaboration graph is 0.22.

We will end this section with a straightforward application of vertex similarity to recommender systems. As the name suggests, recommender systems produce a list of recommendations for the user based on the user's taste, identification of "similar" users, and expert recommendations. For example, Amazon shows their customers items related to what they viewed; Netflix recommends movies based on movies watched; Pandora asks users to "like" songs and plays similar songs; and most news websites provide users with articles similar to the ones they read. The techniques used in recommendation systems can be broadly divided into collaborative filtering techniques and content-based filtering techniques. Content-based filtering techniques use characteristics of the content under consideration and opinions of experts. Collaborative filtering techniques rely on a user's previous choices and all other users who made similar choices.

We present a simple collaborative filtering model where customers and the items they purchase are modeled as a bipartite graph with customers in one vertex class and items in the other. Each customer is linked by edges to items she purchases. Storing the bipartite graph as a list of its vertices together with their neighborhoods quickly gives a list of customers for each item and vice-versa. When a customer c purchases an item i, the similarity $s(c, x)$ for every $x \in N(i)$ is computed. Recommendations for additional items come from the items purchased by vertices with higher values of $s(c, x)$.

For example, consider the customer-item bipartite graph shown in Figure 3.1. Customer c_4 buys item i_3. Since $N(i_3) = \{c_2, c_4, c_5\}$, customer c_4 is compared with c_2 and c_5. From the graph, $N(c_2) = \{i_1, i_3\}$, $N(c_4) = \{i_3, i_4\}$, and $N(c_5) = \{i_1, i_3, i_5\}$. So $s(c_4, c_2) = \frac{1}{3}$ and $s(c_4, c_5) = \frac{1}{4}$. Since $s(c_4, c_2)$ is higher, the other items purchased by c_2 are recommended for c_4.

Collaborative filtering is an example of how a large dataset can be analyzed using mathematical techniques to give useful information without any knowledge of the objects it represents. There are obvious advantages to the seller since sales increase, and in fact sales of less popular items increase because they are brought to the customer's attention. The customer also benefits since he is presented with choices he may not have thought of on his own. On the other hand, this sort of system is also used to present users news articles similar to what they read, often with no indication that a small subset of the daily news is being presented. An echo chamber is an environment where users only encounter information and opinions

Figure 3.1   Customer-item bipartite graph.

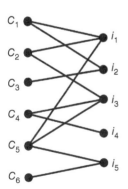

that reinforce or reflect their own opinions. When collaborative filtering is used to recommend content, users may find themselves in echo chambers without even knowing it (Ge et al., 2020).

In conclusion, we seem to have started this section with Granovetter's work that helped people break out of their echo chambers of poverty and violence, and we are ending it with recommender systems that put all of us back in different sorts of echo chambers. Mathematical techniques have the power to help or harm depending on how they are used.

3.2 Centrality Measures

In this section we will look at ways of determining which vertices are more important than other vertices. It was evidently a paradigm shift in sociological thinking to view the importance of an individual as a function of his or her social network. Previously it was assumed to be a function of personal attributes (Bavelas, 1950).

A centrality measure is an assignment of numbers to the vertices of a graph so that they can be ranked. One obvious way of flagging important vertices in a social network is to pick the vertices with high degrees since they know the most people. The degree of a vertex in this context is called its *degree centrality*. If we want to compare the importance of vertices in different networks, then it makes sense to normalize the degree by dividing it by the maximum possible degree. In a graph with n vertices, the maximum degree is $n - 1$. So, the *normalized degree centrality* of a vertex v in a graph with at least two vertices is

$$normdeg(v) = \frac{deg(v)}{n - 1}.$$

The next two centrality measures are based on the shortest paths between two vertices.[4] Recall from Section 2.2, that the total distance of a vertex v in a connected graph is defined as

$$td(v) = \sum_{w \in V} d(v, w).$$

This quantity is a measure of decentrality, so it makes sense to take the reciprocal as a measure of centrality. The *closeness centrality* of vertex v in a connected graph with at least two vertices, denoted by $close(v)$, is defined as

$$close(v) = \frac{1}{td(v)}.$$

It is customary to normalize it by inserting n in the numerator. The *normalized closeness centrality* of vertex v is

$$normclose(v) = \frac{n}{td(v)}.$$

---

4 Linton Freeman's paper "Centrality in networks" has these definitions and the early history of the topic (Freeman, 1979).

Betweenness centrality is based on the frequency with which a vertex falls between pairs of other vertices on the shortest paths connecting them. It originated as a way of quantifying the control of a person on the communication between people in a network. Someone who is strategically located on communication paths linking pairs of people can control the group by withholding or distorting information in transmission.

Let v be a vertex in a connected graph with at least three vertices. Let s_{ab} denote the number of shortest paths between vertices a and b. Let $s_{ab}(v)$ denote the number of shortest paths between a and b that pass through v. The *betweenness centrality* of vertex v, denoted by $bet(v)$, is defined as

$$bet(v) = \sum_{a \neq b \neq v} \frac{s_{ab}(v)}{s_{ab}}.$$

In the following diagram, the shortest path from vertex a to vertex b has length 5 and there are six such shortest paths, of which four 4 pass through vertex v. So $\frac{s_{ab}(v)}{s_{ab}} = \frac{4}{6}$.

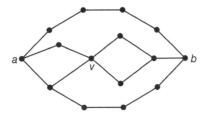

Betweenness centrality can be normalized by dividing it by the number of pairs of distinct vertices $a, b \neq v$. The *normalized betweenness centrality* of vertex v is

$$normbet(v) = \frac{1}{\binom{n-1}{2}} \sum_{a \neq b \neq v} \frac{s_{ab}(v)}{s_{ab}}.$$

For example, consider the graph G in Figure 1.3 redrawn with centrality measures. The degrees of vertices v_1, v_2, v_3, v_4, v_5 are, respectively,

$$2, 2, 3, 2, 3.$$

Observe that

$$td(v_1) = \sum_{w \neq v_1} d(v_1, w) = 1 + 2 + 2 + 1 = 6.$$

Similarly, $td(v_2) = 6, td(v_3) = 5, td(v_4) = 6$, and $td(v_5) = 5$. So closeness centralities for vertices v_1, v_2, v_3, v_4, v_5 are, respectively,

$$\frac{1}{6}, \frac{1}{6}, \frac{1}{5}, \frac{1}{6}, \frac{1}{5}.$$

The betweenness centrality of v_1 is

$$bet(v_1) = \frac{s_{v_2v_3}(v_1)}{s_{v_2v_3}} + \frac{s_{v_2v_4}(v_1)}{s_{v_2v_4}} + \frac{s_{v_2v_5}(v_1)}{s_{v_2v_5}} + \frac{s_{v_3v_4}(v_1)}{s_{v_3v_4}} + \frac{s_{v_3v_5}(v_1)}{s_{v_3v_5}} + \frac{s_{v_4,v_5}(v_1)}{s_{v_4v_5}}$$

$$= 0 + 0 + \frac{1}{2} + 0 + 0 + 0 = \frac{1}{2}.$$

Betweenness centralities for vertices v_1, v_2, v_3, v_4, v_5 are, respectively,

$$\frac{1}{2}, \frac{1}{2}, \frac{3}{2}, 0, \frac{3}{2}.$$

Observe that closeness centrality does not differentiate v_4 from v_1 and from v_2 any better than degree centrality, however, betweenness centrality differentiates them. That said, these measures make more sense for ranking vertices in large graphs.

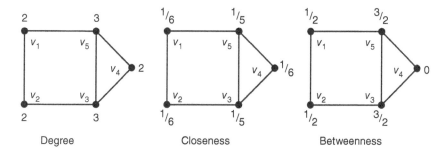

Degree   Closeness   Betweenness

Next, consider the complete graph K_n, for $n \geq 2$. The degree of every vertex is $n - 1$. The total distance of every vertex is $n - 1$, since it is joined by an edge to the other $n - 1$ vertices. Therefore closeness centrality of every vertex is $\frac{1}{n-1}$. The betweenness centrality of every vertex is 0 since every vertex is adjacent to every other vertex.

Finally, consider the star graph $K_{1,n-1}$, for $n \geq 2$ with vertex u in one class and vertices $v_1, \ldots, v_{n-1}$ in the other class. Observe that $deg(u) = n - 1$ and $deg(v_i) = 1$, for $1 \leq i \leq n - 1$. As noted in Section 2.2, $td(u) = n - 1$ and $td(v_i) = 2n - 3$. So

$$close(u) = \frac{1}{n - 1}$$

and

$$close(v_i) = \frac{1}{2n - 3}.$$

Since vertex u is in every path, $bet(u) = \binom{n-1}{2}$ and $bet(v_i) = 0$. Observe that the centrality of u is highest with respect to each of the three centrality measures.

It is often useful to obtain a summary number for the entire graph. Such a number is called a *centralization index*. Freeman notes that a centralization index should capture the extent to which the centrality measure of the most "central" vertex in a graph exceeds the centrality of other vertices. It should be expressed as a ratio of that excess to its maximum possible value for the graph containing the observed number of points (Freeman, 1979). Moreover, Freeman argues that the star graph is the most centralized graph from a practical point of view such as leadership within a network. A centralized network has clearly designated leaders, and most centralized is a network with one leader who oversees all members, none of whom know each other.

Let G be a graph on n vertices and let v^* be the most central vertex based on a specific centrality measure. Freeman defines the *centralization index* as

$$CI(G) = \frac{\sum\limits_{v \in G} \left[centrality(v^*) - centrality(v) \right]}{\sum\limits_{v \in K_{1,n-1}} \left[centrality(v^*) - centrality(v) \right]}.$$

Let us compute the denominator for degree, closeness, and betweenness centrality for $K_{1,n-1}$ with vertex u in one class and vertices $v_1, \ldots, v_{n-1}$ in the other class. We saw earlier that vertex u has the largest value of degree, closeness, and betweenness centrality. Take v^* to be u in the aforementioned definition to obtain:

$$\sum_{i=1}^{n} \left[deg(u) - deg(v_i) \right] = (n-1)\left[n-1-1\right] = (n-1)(n-2),$$

$$\sum_{i=1}^{n} \left[close(u) - close(v_i) \right] = (n-1)\left[\frac{1}{n-1} - \frac{1}{2n-3} \right] = \frac{n-2}{2n-3},$$

and

$$\sum_{i=1}^{n} \left[bet(u) - bet(v_i) \right] = (n-1)\left[\binom{n-1}{2} - 0 \right] = \frac{(n-1)^2(n-2)}{2}.$$

Thus the centralization index of a graph with respect to degree, closeness, and betweenness can be computed by identifying the most central vertex, computing the numerator

$$\sum_{v \in G} \left[centrality(v^*) - centrality(v) \right],$$

and then dividing by the appropriate denominator. In summary

$$CI_{deg}(G) = \frac{1}{(n-1)(n-2)} \sum_{v \in G} \left[deg(v^*) - deg(v) \right],$$

$$CI_{close}(G) = \frac{2n-3}{n-2} \sum_{v \in G} \left[close(v^*) - close(v) \right],$$

and

$$CI_{bet}(G) = \frac{2}{(n-1)^2(n-2)} \sum_{v \in G} \left[bet(v^*) - bet(v) \right].$$

For social networks especially, it helps to compute this index to get an idea of whether the graph has a strong hierarchical structure or whether it is decentralized. A larger centralization index indicates that the graph is more hierarchical.

3.3 Eigenvector and Katz Centrality

The centrality measures in Section 3.2 relied on shortest paths between vertices. The centrality measures in this section and in Section 3.4 rely on walks between vertices.

Let G be a connected graph with n vertices $v_1, \dots, v_n$. Recall from Section 2.4 that the adjacency matrix $A = [a_{ij}]$ is a symmetric matrix with rows and columns labeled by vertices, where each entry a_{ij} is 1 if $v_i v_j$ is an edge and 0 otherwise. The degree of vertex v_i is the row sum, and it may be written as

$$deg(v_i) = \sum_{j=1}^{n} a_{ij}.$$

In 1949 John Seeley proposed an alternative to degree centrality taking into consideration that knowing people who are important makes a person more important (Seeley, 1949). He defined this sort of centrality recursively as

$$centrality(v_i) = \frac{1}{\lambda} \sum_{j=1}^{n} a_{ij} [centrality(v_j)]$$

where λ is a scaling factor. Defined this way, the centrality of v_i is proportional to the centralities of its neighbors.

The *eigenvector centrality* of vertex v_i is defined as the ith entry in $\bar{x}$, where $\bar{x} = \frac{1}{\lambda} A \bar{x}$ and λ is the largest eigenvalue of A. The largest eigenvalue is a unique real number that is greater than 1 and it has an eigenvector with all positive entries.[5]

---

5 This is guaranteed by the Perron–Frobenius Theorem, which states that if M is an irreducible square matrix with non-negative entries, the M has a positive real eigenvalue λ with multiplicity one whose eigenvector has all the positive entries, and if μ is any other eigenvalue of M, then $|\mu| \le \lambda$. In this context a square matrix M is called reducible if, subject to some permutation of the rows and the same permutation of the columns, M can be written in block form as $\begin{pmatrix} B & C \\ O & D \end{pmatrix}$ where O is a zero matrix and B and D are square matrices. A matrix that is not reducible is called irreducible. See Horn and Johnson (1990). It can be shown that the adjacency matrix of a connected graph is irreducible. Furthermore, since the largest eigenvalue λ has a positive eigenvector, $\lambda \ge 1$.

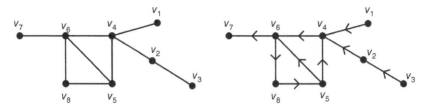

Figure 3.2 A graph and a digraph for centrality measures.

Now if $\bar{x}$ is an eigenvector for A with eigenvalue λ, then $c\bar{x}$ is also an eigenvector for any $c \neq 0$. Therefore we will choose one "representative" eigenvector; specifically with c chosen so that the largest entry in $\bar{x}$ is 1.[6]

For example, the adjacency matrix of the graph in Figure 3.2 is

$$A = \begin{bmatrix} 0 & 0 & 0 & 1 & 0 & 0 & 0 & 0 \\ 0 & 0 & 1 & 1 & 0 & 0 & 0 & 0 \\ 0 & 1 & 0 & 0 & 0 & 0 & 0 & 0 \\ 1 & 1 & 0 & 0 & 1 & 1 & 0 & 0 \\ 0 & 0 & 0 & 1 & 0 & 1 & 0 & 1 \\ 0 & 0 & 0 & 1 & 1 & 0 & 1 & 1 \\ 0 & 0 & 0 & 0 & 0 & 1 & 0 & 0 \\ 0 & 0 & 0 & 0 & 1 & 1 & 0 & 0 \end{bmatrix}$$

The Largest eigenvalue of A is approximately 2.844 with a corresponding eigenvector approximately

$$\bar{x} = \begin{bmatrix} 0.321 \\ 0.366 \\ 0.129 \\ 0.913 \\ 0.909 \\ 1.000 \\ 0.352 \\ 0.671 \end{bmatrix}.$$

This means that the eigenvector centrality of vertex v_1 is 0.321, of vertex v_2 is 0.366, of vertex v_3 is 0.129, and so on. Note that while $deg(v_1) = deg(v_3) = 1$, v_1 has higher eigenvector centrality than v_3. Intuitively, this is because v_1 is connected to v_4, whereas v_3 is connected to v_2, and v_4 has higher degree than v_2.

---

6 There are other methods of choosing a representative eigenvector, including choosing c so that the sum of the entries in $\bar{x}$ is 1.

The obvious advantage of using eigenvector centrality is that if two vertices have roughly the same number of neighbors, then the vertex with more important neighbors will have higher centrality. (In this setting vertices with higher degree are important.) A disadvantage is that if a vertex has few important neighbors, then that vertex could have a lower centrality than a vertex with many unimportant neighbors.

In 1953 Leo Katz proposed a centrality measure taking into consideration that a vertex is important, not only because it has important neighbors, but also because of important neighbors twice removed, and so on (Katz, 1953). Subsequent work by sociologist Philip Bonacich fine-tuned this notion (Bonacich, 1987). Katz added two enhancements to eigenvector centrality:

(1) Instead of scaling A by $\frac{1}{\lambda}$, he scaled A by a constant α, such that $0 < \alpha < \frac{1}{\lambda}$.

(2) He added a constant vector $\overline{\beta}$ to obtain

$$\overline{x} = \alpha A \overline{x} + \overline{\beta}.$$

The previous expression can be simplified as $(I_n - \alpha A)\overline{x} = \overline{\beta}$, where I_n denotes the $n \times n$ identity matrix. Proposition A.1 (i) implies that if λ is an eigenvalue of A, then $c\lambda$ is an eigenvalue for cA, where c is a non-zero constant. So 1 is an eigenvalue of $\frac{1}{\lambda}A$. Since $0 < \alpha < \frac{1}{\lambda}$, the largest eigenvalue of αA is less than 1.

In general, if M is an $n \times n$ matrix whose largest eigenvalue is $|\lambda| \le 1$, then $\sum_{k \ge 0} M^k$ converges, and $\sum_{k \ge 0} M^k = (I_n - M)^{-1}$. The full proof of this statement is in Horn and Johnson (1990); however, a portion of the proof is straightforward. Observe that

$$(I_n - M)(I_n + M + M^2 + \cdots) = (I_n + M + M^2 \cdots) - (M + M^2 + \cdots) = I_n.$$

Therefore

$$(I_n - M)^{-1} = (I_n + M + M^2 + \cdots) = \sum_{k \ge 0} M^k.$$

Using this result, we may conclude that $(I_n - \alpha A)^{-1}$ exists and that $\overline{x} = (I_n - \alpha A)^{-1}\overline{\beta}$.

Let G be a graph with vertices $v_1, \ldots, v_n$ and let A be its adjacency matrix. The *Katz centrality* of vertex v_i is the ith entry in $\overline{x}$, where

$$\overline{x} = (I_n - \alpha A)^{-1}\overline{\beta},$$

λ is the largest eigenvalue of A, $0 < \alpha \le \frac{1}{\lambda}$, and $\overline{\beta}$ is a constant vector. Typically, $\overline{\beta}$ is initialized by assigning a small and equal amount of centrality to each vertex v_i.

Returning to the graph in Figure 3.2, the largest eigenvalue of the adjacency matrix was found to be $\lambda = 2.844$. So

$$\frac{1}{\lambda} \approx \frac{1}{2.844} = 0.352.$$

Taking α to be any number smaller than 0.352, say, $\alpha = 0.25$, and making an arbitrary choice for $\bar{\beta}$, say $\bar{\beta} = \bar{1}_n$, Katz centrality is given by the vector

$$\bar{x} = \begin{bmatrix} 2.040 \\ 2.442 \\ 1.611 \\ 4.159 \\ 3.870 \\ 4.285 \\ 2.071 \\ 3.039 \end{bmatrix}.$$

Comparing eigenvector centrality and Katz centrality, note that the Katz centrality of v_3 is larger relative to the other values as compared with its eigenvector centrality. This is because it is assigned weight by the Katz centrality formula that is independent of the vertices that link to it.

Eigenvector and Katz centralities are global measures in the sense that the entire adjacency matrix A must be obtained. This is difficult when the graph is large. For Katz centrality we may apply the approximation derived earlier to avoid using the entire adjacency matrix. The expression $\bar{x} = (I_n - \alpha A)^{-1}$ can be simplified as follows:

$$\bar{x} = (I_n - \alpha A)^{-1}\bar{\beta} = \left[\sum_{k \geq 0}(\alpha A)^k\right]\bar{\beta} = \left[\sum_{k \geq 0}\alpha^k A^k\right]\bar{\beta}.$$

Proposition 2.4.1 implies that the ijth entry in A^k gives the number of $v_i - v_j$ walks of length k. Thus Katz centrality relies on the centrality of neighbors, and neighbors of neighbors, and so on. The presence of α regulates the rate at which the contribution of walks decreases as their length increases. It may be viewed as an "attenuation factor" or "damping factor" to indicate that the importance of neighbors of neighbors should weigh less than the importance of immediate neighbors. Since $0 < \alpha \leq 1$, α^k is highest for immediate neighbors and gets rapidly smaller thereafter. Practically speaking the calculation of $\sum \alpha^k A^k$ may be stopped at a small value of k since the contribution of higher values of α^k will be negligible. In conclusion, the Katz centrality of vertex v_i is the ith entry in $\bar{x}$, where

$$\bar{x} = \left[\sum_{k \geq 0}\alpha^k A^k\right]\bar{\beta}$$

and we can approximate the values or $\bar{x}$ by computing the values of the series up to a fixed value of k. For example, using $\alpha = 0.25$, $\bar{\beta} = \bar{1}_n$, and stopping at $k = 3$, we get

$$\bar{x} = [I_n + \alpha A + \alpha^2 A^2 + \alpha^3 A^3]\bar{\beta} = \begin{bmatrix} 1.500 \\ 1.812 \\ 1.375 \\ 2.625 \\ 2.375 \\ 2.625 \\ 1.500 \\ 1.938 \end{bmatrix}.$$

Note that even with $k = 3$, the values are already proportional to the actual values obtained earlier. However, there are some differences; in particular v_6 has a higher centrality than v_4, but in the approximation the values for both are equal.

The advantage of computing $\bar{x}$ using this series is that computing $A^{k+1} = A \times A^k$ can be done without storing all of A or A^k in memory at once since the ijth entry of A^{k+1} depends only on the ith row of A and the jth column of A^k. This is relevant for large graphs, and we will see in Section 3.4 that it is a key part of the usefulness of PageRank.

Next, let us look at eigenvector and Katz centrality for digraphs. If G is a directed graph with adjacency matrix $A = [a_{ij}]$, $outdeg(v_i)$ is the sum of ones in row v_i and $indeg(v_i)$ is the sum of ones in column v_i. The adjacency matrix of the digraph in Figure 3.2 is

$$A = \begin{array}{c} \\ v_1 \\ v_2 \\ v_3 \\ v_4 \\ v_5 \\ v_6 \\ v_7 \\ v_8 \end{array} \begin{array}{cccccccc} v_1 & v_2 & v_3 & v_4 & v_5 & v_6 & v_7 & v_8 \\ \begin{pmatrix} 0 & 0 & 0 & 1 & 0 & 0 & 0 & 0 \\ 0 & 0 & 0 & 1 & 0 & 0 & 0 & 0 \\ 0 & 1 & 0 & 0 & 0 & 0 & 0 & 0 \\ 0 & 0 & 0 & 0 & 0 & 1 & 0 & 0 \\ 0 & 0 & 0 & 1 & 0 & 1 & 0 & 0 \\ 0 & 0 & 0 & 0 & 0 & 0 & 1 & 1 \\ 0 & 0 & 0 & 0 & 0 & 0 & 0 & 0 \\ 0 & 0 & 0 & 0 & 1 & 0 & 0 & 0 \end{pmatrix} \end{array}.$$

If we are interested in a centrality measure based on outdegrees, then the situation is exactly as for graphs. Suppose we are interested in indegrees. In the Web digraph, for example, vertices are webpages and the arcs are hypertext links. A webpage is important if many pages point to it. Therefore the indegree matters more than the outdegree in this setting. In this case it makes sense to work with the transpose of the adjacency matrix shown in the following text since the indegrees are row sums in A^T:

$$A^T = \begin{array}{c} \\ v_1 \\ v_2 \\ v_3 \\ v_4 \\ v_5 \\ v_6 \\ v_7 \\ v_8 \end{array} \begin{array}{cccccccc} v_1 & v_2 & v_3 & v_4 & v_5 & v_6 & v_7 & v_8 \\ \left(\begin{array}{cccccccc} 0 & 0 & 0 & 0 & 0 & 0 & 0 & 0 \\ 0 & 0 & 1 & 0 & 0 & 0 & 0 & 0 \\ 0 & 0 & 0 & 0 & 0 & 0 & 0 & 0 \\ 1 & 1 & 0 & 0 & 1 & 0 & 0 & 0 \\ 0 & 0 & 0 & 0 & 0 & 0 & 0 & 1 \\ 0 & 0 & 0 & 1 & 1 & 0 & 0 & 0 \\ 0 & 0 & 0 & 0 & 0 & 1 & 0 & 0 \\ 0 & 0 & 0 & 0 & 0 & 1 & 0 & 0 \end{array}\right) \end{array}.$$

Hence if a_{ij}^T is the ijth entry of A^T, then

$$indeg(v_i) = \sum_{j=1}^n a_{ij}^T.$$

Just as with the degree sequence, the indegree sequence may be viewed as a vector $\bar{x}$ and

$$A^T \bar{1}_n = \bar{x}.$$

Eigenvector centrality with respect to $indegree(v_i)$ is the ith entry in $\bar{x}$, where

$$\bar{x} = \frac{1}{\lambda} A^T \bar{x},$$

and λ is the largest eigenvalue of A^T. Note that by Proposition A.1 (iii) A and A^T have the same eigenvalues.

Katz centrality with respect to indegree of vertex v_i is the ith entry in $\bar{x}$, where

$$\bar{x} = \alpha A^T \bar{x} + \bar{\beta} = \left[\sum_{k \geq 0} (\alpha A^T)^k\right] \bar{\beta} = \left[\sum_{k \geq 0} \alpha^k (A^T)^k\right] \bar{\beta}.$$

In the previous expression λ is the largest eigenvalue of A, $0 < \alpha \leq \frac{1}{\lambda}$, and $\bar{\beta}$ is a constant vector.

For the digraph in Figure 3.2 the largest eigenvalue of A is 1.221, and it has eigenvector

$$\begin{bmatrix} 0 \\ 0 \\ 0 \\ 0.550 \\ 0.671 \\ 1.000 \\ 0.819 \\ 0.819 \end{bmatrix}.$$

Katz centrality with $\alpha = 0.25$ and $\beta = \bar{1}$ is

$$\begin{bmatrix} 1.000 \\ 1.250 \\ 1.000 \\ 1.903 \\ 1.364 \\ 1.817 \\ 1.454 \\ 1.454 \end{bmatrix}.$$

Since $\bar{\beta}$ is the vector of ones, all the vertices are assigned some centrality even if no other vertices link to them. Finally, using the approximation with $k = 3$ gives a slightly different vector:

$$\begin{bmatrix} 1.000 \\ 1.250 \\ 1.000 \\ 1.875 \\ 1.312 \\ 1.750 \\ 1.375 \\ 1.375 \end{bmatrix}.$$

Comparing eigenvector and Katz centrality, observe that vertex v_2 has zero eigenvector centrality even though it has positive indegree, because the only vertex that links to it is v_3 and v_3 has zero eigenvector centrality. However, v_2 has greater Katz centrality, including a small contribution due to the link from v_3.

3.4 PageRank

PageRank is Larry Page and Sergey Brin's algorithm that launched Google in 1998 (Brin and Page, 1998). It is a modification of Katz centrality. PageRank is based on the premise that in the Web digraph a vertex is important because of other important and "frugal" vertices that link to it directly or indirectly or if it is highly linked by frugal vertices. In this context a frugal vertex means a vertex with a small outdegree. An important webpage that is also generous with links does not contribute as much as an important webpage with few links. According to PageRank's premise, each vertex has an equal "budget" of importance that it distributes to the other vertices to which it links.

Consider two vertices x and y in a digraph and suppose $outdeg(x) = 25$ and $outdeg(y) = 100$. Using Katz centrality measure, every outgoing arc contributes equally to the centrality of the vertex it points to. Therefore vertex y will contribute a total of four times more centrality to the vertices it points to than vertex x. PageRank addresses this issue by dividing the contribution each vertex makes by its outdegree.

Let G be a digraph with adjacency matrix $A = [a_{ij}]$. Let C be the $n \times n$ matrix obtained from A by dividing each entry in a row by the sum of all the entries in the row (the outdegree).

$$c_{ij} = \frac{a_{ij}}{\sum_{k=1}^{n} a_{ik}}.$$

So the sum of each row in C is 1. The next result establishes that 1 is the largest eigenvalue of C.

Proposition 3.4.1. *Let C be an $n \times n$ matrix with non-negative real entries such that the sum of the entries in each row is 1. Then the largest eigenvalue of C is 1.*

Proof. Since each row sum of C is 1, $C\overline{1}_n = \overline{1}_n$ and therefore 1 is an eigenvalue of C. Suppose λ is another eigenvalue of C with eigenvector

$$\overline{x} = \begin{bmatrix} x_1 \\ x_2 \\ \vdots \\ x_n \end{bmatrix}.$$

Then $C\overline{x} = \lambda\overline{x}$; that is,

$$C\overline{x} = \begin{bmatrix} \sum_j c_{1j}x_j \\ \sum_j c_{2j}x_j \\ \vdots \\ \sum_j c_{ij}x_j \\ \vdots \\ \sum_j c_{nj}x_j \end{bmatrix} = \begin{bmatrix} \lambda x_1 \\ \lambda x_2 \\ \vdots \\ \lambda x_i \\ \vdots \\ \lambda x_n \end{bmatrix}.$$

Let x_i denote the largest value in $\overline{x}$. Then

$$\lambda x_i = \sum_j c_{ij}x_j \leq \sum_j c_{ij}x_i = x_i \sum_j c_{ij} = x_i(1) = x_i.$$

Therefore $\lambda \leq 1$. $\qquad\qquad\qquad\qquad\qquad\qquad\qquad\qquad\qquad\qquad\qquad\square$

The *PageRank* of vertex v_i is defined as the ith entry in $\overline{x}$, where

$$\overline{x} = \alpha C^T \overline{x} + \overline{\beta} = \left[\sum_{k \geq 0} \alpha^k (C^T)^k \right] \overline{\beta}, \text{ where } 0 < \alpha < 1$$

Observe that in the definition of PageRank, there is no mention of eigenvalues or eigenvectors.

For the digraph in Figure 3.2

$$
C^T = \begin{array}{c} \\ v_1 \\ v_2 \\ v_3 \\ v_4 \\ v_5 \\ v_6 \\ v_7 \\ v_8 \end{array}
\begin{array}{cccccccc}
v_1 & v_2 & v_3 & v_4 & v_5 & v_6 & v_7 & v_8
\end{array}
\left(\begin{array}{cccccccc}
0 & 0 & 0 & 0 & 0 & 0 & 0 & 0 \\
0 & 0 & 1 & 0 & 0 & 0 & 0 & 0 \\
0 & 0 & 0 & 0 & 0 & 0 & 0 & 0 \\
1 & 1 & 0 & 0 & \frac{1}{2} & 0 & 0 & 0 \\
0 & 0 & 0 & 0 & 0 & 0 & 0 & 1 \\
0 & 0 & 0 & 1 & \frac{1}{2} & 0 & 0 & 0 \\
0 & 0 & 0 & 0 & 0 & \frac{1}{2} & 0 & 0 \\
0 & 0 & 0 & 0 & 0 & \frac{1}{2} & 0 & 0
\end{array}\right)
$$

and the PageRank is

$$
\bar{x} = \begin{bmatrix} 0.097 \\ 0.122 \\ 0.097 \\ 0.168 \\ 0.127 \\ 0.155 \\ 0.117 \\ 0.117 \end{bmatrix}.
$$

The vector obtained with $k = 3$ iterations of the sum is

$$
\begin{bmatrix} 0.098 \\ 0.123 \\ 0.098 \\ 0.169 \\ 0.126 \\ 0.154 \\ 0.115 \\ 0.115 \end{bmatrix}.
$$

For the example graph, PageRank converges much more quickly than Katz centrality.

Walk-based centrality measures like eigenvector, Katz, and PageRank technically apply only to digraphs with at least one cycle. This is because if the directed graph has no cycles, then for some k there are no paths of length k, and therefore $A^k = 0$, where A is the adjacency matrix.

In the original 1998 paper Brin and Page defined PageRank as follows:

We assume page A has pages $T1, \ldots, Tn$, which point to it (i.e. are citations). The parameter d is a damping factor which can be set between 0 and 1. We usually set d to 0.85 ... Also $C(A)$ is defined as the number of links going out of page A. The PageRank of a page A is given as follows:

$$PR(A) = (1 - d) + d \left(\frac{PR(T1)}{C(T1)} + \cdots + \frac{PR(Tn)}{C(Tn)} \right).$$

The notation $PR(A)$ in the previous paragraph stands for PageRank of A. In the terminology we use, let $v_1, \ldots, v_n$ represent the vertices of the Web digraph and let $A = [a_{ij}]$ be its adjacency matrix. Let $v_{i_1}, \ldots, v_{i_p}$ be the vertices pointing to v_i. Then

$$PR(v_i) = (1 - d) + d \left(\frac{PR(v_{i_1})}{outdeg(v_{i_1})} + \cdots + \frac{PR(v_{i_p})}{outdeg(v_{i_p})} \right).$$

Note that column i in the adjacency matrix has a value of 1 corresponding to vertices pointing to v_i and 0 everywhere else; that is, $a_{ki} = 1$ if vertex v_k points to v_i and $a_{ki} = 0$ otherwise. If we substitute α for d, and let $\bar{\beta}$ be a constant vector whose values are all $1 - d$, then since the ijth position in C^T is $\frac{a_{ji}}{\sum\limits_{k=1}^{n} a_{jk}}$, we can rewrite the expression for $PR(v_i)$ as

$$PR(v_i) = (1 - d) + d \left(\frac{a_{1i}}{\sum\limits_{k=1}^{n} a_{1k}} PR(v_1) + \cdots + \frac{a_{ni}}{\sum\limits_{k=1}^{n} a_{nk}} PR(v_n) \right).$$

This is the ith entry in the vector $\bar{\beta} + \alpha C^T \bar{x}$.

It is important to see that $PR(v_i)$ can be computed using only $PR(v_j)$ for vertices v_j that point to v_i, together with the ith row of C^T, which contains nonzero values only for the other vertices pointing to v_i; no other information about the graph is needed. Thus for a very large graph, updates to PR can be computed using a distributed network of computers and updates to PR values resulting from changes in the graph can be computed efficiently.

Brin and Page began by properly modeling the system they were studying, the web digraph. Typical user behavior on the web is to either access a page via a link from another page, or directly type a URL in the browser (or use a bookmark). The formulas express this process in mathematical terms. In particular, d is the probability that a user arrives at a page via a link and $1 - d$ is the probability that a user arrives at a page directly. A good value of d is 0.85, they say, which means that when a user switches to a new page, she does so by clicking on a link 85% of the time and by visiting the page directly 15% of the time.

In conclusion, PageRank can be explained with no mention of eigenvalues. It does, however, help to see it in the context of the other walk-based centrality measures and know that there are some deep results behind it.

Exercises

3.1 For the graph in Figure 3.3:
  (a) Find the neighborhood similarity of pairs of vertices and identify similar vertices based on this measure (use a pre-determined criterion you set).
  (b) Identify strong and weak ties.
  (c) Identify local bridges if any, and find their spans.
  (d) Find the local clustering coefficient for each vertex.
  (e) Find the mean local clustering coefficient and global clustering coefficient.
  (f) Find degree, closeness, and betweenness centralities.
  (g) Find eigenvector centrality.
  (h) Find Katz centrality.

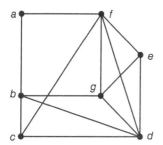 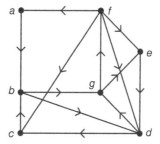

Figure 3.3 A graph and a digraph.

3.2 For the strongly connected digraph in Figure 3.3:
  (a) Find indegree centrality and outdegree centrality.
  (b) Find closeness and betweenness centrality by applying the definitions of closeness and betweenness centrality of graphs to digraphs. (Replace distance between a pair of vertices with the directed distance.)
  (c) Find Katz centrality.
  (d) Find PageRank.

3.3 The graphs in Figure 3.4 are taken from Schoch and Brandes (2016). Calculate the degree, closeness, betweenness, and eigenvector centrality. (These graphs are interesting because these invariants give consistent rankings for the first graph, but vary considerably for the second graph.)

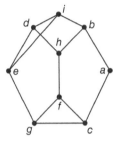 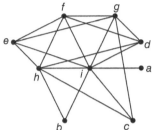

Figure 3.4 Schoch and Brandes graphs.

3.4 Consider the map shown in Figure 3.5. Construct a graph by defining vertices as intersections of streets and edges as the streets themselves. Assume the graph is undirected. Determine the most important vertices using the different centrality measures. Interpret your results.

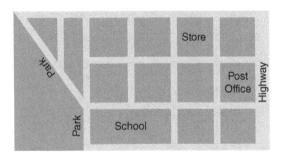

Figure 3.5 A small town map.

3.5 Invariants in this chapter are meaningful for large graphs. Many datasets are available online. From the toy examples in Chapters 1 and 2, step up to larger graphs and use software for calculations. With every order of magnitude increase, calculations become harder and complications arise. Find the path-based and walk-based centrality measures for the following large graphs.

(a) The karate club network, a social network of 34 members of a karate club (Zachary, 1977).

(b) The social network of 62 dolphins that were living off the coast of New Zealand (Lusseau et al., 2003).

(c) The social network of 215 elite Florentine families used to study the Medici family's influence and control (Padgett and Ansell, 1993).

(d) A portion of the connectome of the cat brain with 65 nodes.[7]

(e) The Fly Medulla Connectome with 1781 vertices and 33 641 edges.

Topics for Deeper Study

3.6 Hyperlink-Induced Topic Search (HITS) is Jon Kleinberg's link-analysis algorithm for the Web (Kleinberg, 1999). Like PageRank, it is designed for the Web digraph. This algorithm identifies authoritative pages, which are linked to by many other pages, and hub pages, which link to many authorities. Like Katz and PageRank centralities, HITS can be thought of as ranking pages using a method that requires finding eigenvalues and eigenvectors of a matrix. Specifically, authority scores are entries in an eigenvector associated with the largest eigenvalue of $A^T A$, and hub scores are entries in an eigenvector associated with the largest eigenvalue of AA^T, where A is the adjacency matrix of the digraph. Apply Kleinberg's algorithm to the digraph in Figure 3.2. Which vertices does the algorithm indicate as hubs, and which are authorities? Compare the results with Katz and PageRank centrality values for the same digraph, and offer an explanation for the differences.

3.7 Find the strong and weak ties and central nodes in the transportation network of a small town modeled as an undirected graph. Include a description of the software used and any discrepancies that arise when using different software packages.

3.8 Collaborative filtering systems suffer from a defect known as the "cold start" problem. Namely, no recommendations can be made to a brand-new customer, and when a new item is added to the system it will never be recommended. The solution to this problem is to combine content-based and collaborative techniques, so that recommendations are based on common customer and product attributes in addition to purchase histories. Review the literature in this area and compare alternate approaches.

3.9 The MovieLens dataset (Harper and Konstan, 2015) is a popular and growing dataset for evaluating recommender systems. Download the current version of the dataset and apply a simple collaborative filtering approach.

---

7 Connectomes are located at https://neurodata.io/project/connectomes/

3.10 Wikipedia makes its data easily available. Pick a field of study and determine the most central topics in it by constructing a digraph with Wikipedia pages as vertices. Put an arc between pages u and v if page u has a hypertext link to page v.

3.11 Turn each Shakespearean play (or any other play or movie) into an undirected weighted graph, where two characters are adjacent if they appear in the same scene, counting the number of scenes they appear in as the weight on the edge. Determine the most central characters using centrality measures customized for weighted graphs.

4

Types of Networks

This chapter describes some of the popular types of networks such as scale-free, small-world, assortative, and covert networks. In each case the adjective describing the term "network" applies to a different aspect of the network. Small-world networks in Section 4.1 are networks that have certain properties markedly different from random graphs. Scale-free networks in Section 4.2 are networks whose degree distributions can be modeled by the power-law function. Assortative mixing in Section 4.3 is a measure of how likely vertices that are similar are linked together and covert networks in Section 4.4 have multiple layers and hidden links.

4.1 Small-World Networks

Erdős and Rényi (1959) and independently Gilbert (1959) used ideas from probability theory to define what they called a "random graph." See Appendix B for a review of elementary probability theory. A *random graph* on n vertices, denoted by $G(n, p)$, is a graph in which each edge occurs independently with probability p, where $0 < p < 1$. To construct a random graph, select a pair of vertices and generate a random number x between 0 and 1. If $x \geq p$, join the vertices by an edge, otherwise let the vertices remain non-adjacent. The probability of getting a graph with m edges is given by the binomial probability distribution

$$B\left(m; \binom{n}{2}, p\right) = \binom{\binom{n}{2}}{m} p^m (1-p)^{\binom{n}{2}-m}.$$

The mean of this distribution is $\binom{n}{2} p$ and the variance is $\binom{n}{2} p(1-p)$.

For example, we can construct a random graph on n vertices by tossing an unbiased coin for each pair of vertices. Heads corresponds to an edge and tails corresponds to the absence of an edge. In this case $p = 0.5$. The graph G with 5 vertices

Graphs and Networks, First Edition. S. R. Kingan.
© 2022 John Wiley & Sons, Inc. Published 2022 by John Wiley & Sons, Inc.

and 6 edges in Figure 1.3 shows an instance of the random graph $\mathcal{G}(5, 0.5)$. Tossing a coin for the pair of vertices $v_1 v_2$ resulted in a head, so the two vertices are joined by an edge. Tossing a coin for the pair of vertices $v_1 v_3$ resulted in a tail, so there is no edge between them, and so on for each of the 10 possible pairs of vertices

$$v_1 v_2, v_1 v_3, v_1 v_4, v_1 v_5, v_2 v_3, v_2 v_4, v_2 v_5, v_3 v_4, v_3 v_5, v_4 v_5.$$

So the graph G in Figure 1.3 shown again below, with solid lines representing edges selected and dashed lines representing edges rejected, corresponds to the outcomes

HTTHHTTHHH.

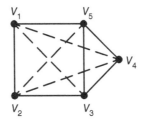

Given $n = 5$, the probability of getting a graph with 6 edges is

$$B(6; 10, 0.5) = \binom{10}{6} \left(\frac{1}{2}\right)^6 \left(\frac{1}{2}\right)^4 = \binom{10}{6} \frac{1}{2^{10}} = \frac{210}{1024} \approx 0.21.$$

We need three properties of random graphs.

(1) The probability that vertex v has degree k follows the binomial distribution

$$B(k; n-1, p) = \binom{n-1}{k} p^k (1-p)^{n-1-k}.$$

In this case the mean degree is $(n-1)p$ and since n is typically very large we may approximate $n-1$ by n. So

$$d(\mathcal{G}(n, p)) = np.$$

(2) Since the probability that two neighbors of a vertex are adjacent is the probability that two random vertices are adjacent, the mean clustering coefficient is approximately p. So

$$C(\mathcal{G}(n, p)) \sim p.$$

(3) The approximate mean distance of a random graph $G(n, p)$ is

$$\mu(\mathcal{G}(n, p)) \sim \frac{ln(n)}{ln(np)}.$$

In 1998, Duncan Watts and Steven Strogatz described a graph as a small-world network if it has a "small" mean distance like a random graph and a "large" clustering coefficient unlike a random graph. They had noticed that many real-world graphs were highly clustered like regular graphs (Watts and Strogatz, 1998). To conclude that a graph G is a small-world network, they compared G with the corresponding random graph $\mathcal{G}(n,p)$ with n vertices and $p = \frac{d(G)}{n}$ to see if $\mu(G)$ was roughly the same as $\mu(\mathcal{G}(n,p))$ and $C(G)$ was much greater than $C(\mathcal{G}(n,p))$.

A graph G with n vertices, mean degree $d(G)$, mean distance $\mu(G)$, and mean local clustering coefficient $C(G)$ is called a *small-world network* if

$$\mu(G) \sim \frac{ln(n)}{ln(np)}$$

and

$$C(G) \gg \frac{d(G)}{n}.$$

Watts and Strogatz concluded that the following three graphs are small-world networks:

(1) The first graph they considered was the film actors graph consisting of 225 226 film actors, where two actors are joined by an edge if they acted in a movie together. They noted that roughly 90% of all the actors, listed in the Internet Movie Database, as of April 1997, were in one connected component and they considered this component as the graph of 225 226 actors. According to their calculations, this graph has mean degree $d(G) = 61$, mean distance

$$\mu(G) = 3.65,$$

and mean local clustering coefficient

$$C(G) = 0.79.$$

The corresponding random graph $\mathcal{G}(n,p)$ with $n = 225\ 226$ and

$$p = \frac{d(G)}{n} = \frac{61}{225\ 226} = 0.000\ 27$$

has mean distance

$$\mu(\mathcal{G}(n,p)) = \frac{ln(n)}{ln(np)} = \frac{ln(225\ 226)}{ln(61)} \sim 2.99$$

and mean clustering coefficient

$$C(\mathcal{G}(n,p)) = 0.000\ 27.$$

Observe that 3.65 is close enough to 2.99 and 0.79 is very much larger than 0.000 27. Since $\mu(G) \sim \mu(\mathcal{G}(n,p))$ and $C(G) \gg C(\mathcal{G}(n,p))$, the actors graph may be viewed as a small-world network.

(2) The next graph was the power grid graph with vertices representing 4941 generators, transformers and substations, where two vertices are adjacent if they are joined by a high voltage transmission line. According to their calculations this graph has $n = 4941$, $d(G) = 2.67$, $C(G) = 0.08$, and $\mu(G) = 18.7$. The random graph $\mathcal{G}(n, p)$ with $n = 4941$ and

$$p = \frac{d(G)}{n} = \frac{2.67}{4941} = 0.005$$

has mean distance $\mu(\mathcal{G}(n, p)) = 12.4$ and mean clustering coefficient $C(\mathcal{G}(n, p)) = 0.005$. Here $18.7 \sim 12.4$ and $0.08 \gg 0.005$. Therefore G may be viewed as a small-world network.

(3) The third graph they considered was the connectome of *C. elegans* described in Section 1.1. As the authors say "we treat all edges as undirected and unweighted, and all vertices as identical, recognizing that these are crude approximations." They took $n = 282$ and obtained $d(G) = 14$, $C(G) = 0.28$, and $\mu(G) = 2.65$. The corresponding random graph $\mathcal{G}(n, p)$ with $n = 282$ and

$$p = \frac{14}{282} = 0.05$$

has mean distance $\mu(\mathcal{G}(n, p)) = 2.25$ and mean clustering coefficient $C(\mathcal{G}(n, p)) = 0.05$. Here $2.65 \sim 2.25$ and $0.28 \gg 0.05$. Therefore G may be viewed as a small-world network.

Since the Erdős–Rényi random graph model does not capture the large clustering coefficient present in real-world graphs, Watts and Strogatz created an alternative model called the Watts–Strogatz model. Consider a graph with n vertices arranged in a ring, where each vertex is joined by an edge to its k nearest neighbors. This graph will have $\frac{nk}{2}$ edges. For each vertex v select the edge joining its nearest neighbor in a clockwise manner. With probability p delete the edge and connect v to a vertex chosen uniformly at random over the entire ring. Duplicate edges are forbidden. Repeat this process by moving clockwise around the ring, considering each vertex in turn until one lap is completed. Next, consider the edges that connect vertices to their second-nearest neighbors clockwise. As before, randomly "rewire" one edge per vertex with probability p. Continue this process, circulating around the ring and proceeding outward to more distant neighbors after each lap, until each edge in the original graph has been considered once. Since G has $\frac{nk}{2}$ edges, the rewiring process stops after $\frac{k}{2}$ laps. By experimenting with different values of n and p, they found a zone between $p = 0$ and $p = 1$ for which the mean clustering coefficient is high and the mean distance is low. In other words they found that for some values of p, the graph has the two small-world properties: a high mean clustering coefficient like a regular graph and a low mean distance like a random graph.

We will end this section with another example of a small-world graph. Let G be the Erdős-1 collaboration graph of the 511 authors with Erdős number 1 (shown in Figure 1.19). For this graph $d(G) = 6.509$, $\mu(G) = 3.807$, and $C(G) = 0.342$. The corresponding random graph $\mathscr{G}(n,p)$ with $n = 511$ and

$$p = \frac{d(G)}{n} = \frac{6.509}{511} = 0.0127$$

has mean distance

$$\mu(\mathscr{G}(n,p)) = \frac{ln(n)}{ln(np)} = \frac{ln(511)}{ln(6.509)} \sim 3.329$$

and clustering coefficient

$$C(\mathscr{G}(n,p)) = p = 0.0127.$$

Since $\mu(G) \sim \mu(G(n,p))$ and $C(G) \gg C(G(n,p))$, this graph is a small-world network.

Graphs serve as models for real-world systems consisting of different types of objects linked together in some manner. The physicist's point of view of networks and reasons why they study them is captured in the following paragraph from Barabási and Albert (1999).

> The inability of contemporary science to describe systems composed of nonidentical elements that have diverse and non-local interactions currently limits advances in many disciplines, ranging from molecular biology to computer science. The difficulty of describing these systems lies partly in their topology: Many of them form rather complex networks whose vertices are the elements of the system and whose edges represent the interactions between them.

4.2   Scale-Free Networks

In Section 1.2 we saw ways of classifying large graphs based on their appearance. Another approach to large graph classification is based on the shape of their degree distributions. Recall from Section 1.1 that scale-free networks are graphs whose degree distributions can be modeled by the power law function

$$y = cx^{-k},$$

where c and k are non-negative constants and $k \geq 1$. Then $\log y$ and $\log x$ have the following linear relationship:

$$\log y = \log cx^{-k} = \log c + (-k)\log x.$$

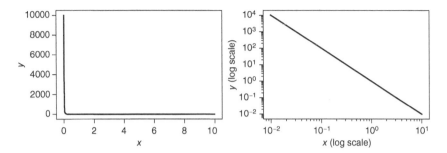

Figure 4.1 $y = \frac{1}{x^2}$ with a standard scale and a log–log scale.

The log-log plot of a power law function (called its signature) is a straight line. For example, the function $f(x) = \frac{1}{x^2}$ is shown in Figure 4.1. The second graph has a log–log scale and shows the signature straight line.

The power law function is not a probability function. However, it can be turned into a probability function by including a suitable constant. Specifically, it can be shown that

$$f(x) = \frac{k-1}{x^k}$$

with domain $[1, \infty)$ and $k \geq 2$ is a probability function since

$$\int_1^\infty \frac{k-1}{x^k} dx = 1.$$

This probability function is called the *Pareto distribution* after the Italian economist Vilfredo Pareto who used it to model income distribution. Pareto observed that 80% of the land in Italy was owned by 20% of the population (Pareto, 1964). This observation that 80% of "effects" come from 20% of "causes" appears in a variety of situations and is called the "80 – 20 rule."

In 1935, linguist George Kingsley Zipf observed that in a corpus of words (this book, for instance) the frequency of a word is inversely proportional to its rank in a word-frequency table (Zipf, 1935).[1] For example, the Brown University Corpus compiled in 1961 has 500 samples of text totaling over a million words. The word "the" is the most frequent word and its frequency is roughly 7%, followed by "of" with frequency roughly 3.5%, "and" with frequency roughly 2.4%, and so on. If c is the frequency of the word that occurs the most, then the distribution of the remaining words follow the power law function $y = \frac{c}{x}$.

In 1999, Albert-László Barabási and Réka Albert published a landmark paper titled "Emergence of scaling in random networks," where they showed that several graphs have scale-free degree distributions (Barabási and Albert, 1999), including the three large graphs presented in Watts and Strogatz (1998), the actors network, the power grid, and the *C. elegans* connectome, as well as several social

---

1 Much has been written about power laws, Pareto distributions, and Zipf's law and the astonishing array of real-world situations they appear to model (Powers, 1998; Newman, 2005).

networks and the Internet graph.[2] Recall from Section 1.1 that a key feature of scale-free networks is "preferential attachment," which means few vertices have high degree (hubs) and most have small degree. The presence of hubs makes scale-free networks highly vulnerable to targeted attacks on the hubs, yet very resilient to random attacks since most vertices have small degree (Albert et al., 2000). Such networks are fragile with respect to the removal of the few high degree vertices because the diameter of the graph increases rapidly when high degree vertices are deleted. They created another model called the Barabási–Albert model, where new vertices were linked to vertices with high degree preferentially and showed that the resulting model was scale-free.

Another example of a scale-free network is a protein-protein interaction network or protein network in short. Each protein is represented by a vertex and the physical interactions between proteins are the edges. In Jeong et al. (2001) the authors created the protein network of the yeast *Saccharomyces cerevisiae*. This graph has 1870 vertices (proteins) and 2240 edges and 93% of the vertices have degree at most 5. The random removal of proteins in a protein network corresponds to random mutations in the genome. The authors concluded that *S. cerevisiae* is resilient to random mutations. In Rain et al. (2001) and Giot et al. (2003), the authors observed a similar architecture for the protein network of the bacterium *Helicobacter pylori* and the insect *Drosophila melanogaster*, respectively.

Consider the citation digraph of high energy physics papers published in the arXiv e-print system with 34 546 vertices (papers) and 421 578 arcs, where there is an arc from vertex x to vertex y if x cites y.[3] The in-degree distribution is shown in Figure 4.2. Observe how the log scale on the y-axis in the second figure

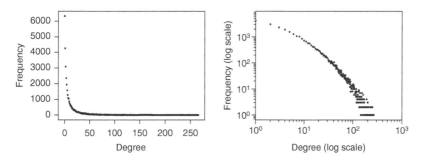

Figure 4.2 The in-degree distribution of the high energy physics citation digraph.

---

2 The scale-free nature of an artificially engineered network like the Internet has been disputed. The reader may find it interesting to read an article that appeared in the *Notices of the AMS* titled "Mathematics and the Internet: A source of confusion and enormous potential" (Willinger et al., 2009) not just for the controversy promised in the title, but also to see the difficulties that arise when working with large graphs.

3 Data for this citation diagraph is available at https://snap.stanford.edu/data/cit-HepPh.html. The downloaded data consists of a text file with a short hender, followed by a list of 421,578 pairs of numbers. Each number identifies one document, and a value (x, y) in the data indicates that paper x cites paper y. There is no information on whether or not a paper cites or is cited by a paper outside the dataset.

reveals information hidden in the long tail of the first figure. Real-world data is not likely to fit a curve exactly and in this case the log–log plot is close enough to a straight line.

4.3 Assortative Mixing

The term assortative refers to people marrying people with similar characteristics. In a graph, *assortative mixing* is the tendency of individuals to connect with others like them. The tendency of individuals to connect with others dierent from them is called *disassortative*. If the graph is not assortative or disassortative, it is called *non-assortative*. Assortativity (also called homophily) applied to networks originated in Newman (2002) and Adamic and Adar (2003). In this section we will present three approaches to assortativity.

Some large graphs have vertices that are of just a few different types. For example, vertices in a social network may be classified according to the racial groups recognized by the Census (White, Black, Asian, Native American and Native Alaskan, Native Hawaiian and Pacific Islander). Vertices in a food web may be classified as plants, herbivores, carnivores, and omnivores. Vertices in a connectome may be grouped based on function or proximity. The physical Internet has several different types of vertices, including providers who run the Internet backbone, Internet Service Providers (ISPs), and end users.

The first table in Figure 4.3 shows the results of a study of 1958 couples Newman (2003b). The second table shows the same data converted into a joint probability distribution, where each entry p_{ij} is the original number divided by the total number of people. The marginal probability distributions a_i (row sum) and b_j (column

		Women			
		Black	Hispanic	White	Other
Men	Black	506	32	69	26
	Hispanic	23	308	114	38
	White	26	46	599	68
	Other	10	14	47	32

		Women				
		Black	Hispanic	White	Other	a_i
Men	Black	0.258	0.016	0.035	0.013	0.323
	Hispanic	0.012	0.157	0.058	0.019	0.247
	White	0.013	0.023	0.306	0.035	0.377
	Other	0.005	0.007	0.024	0.016	0.053
	b_j	0.289	0.204	0.423	0.084	

Figure 4.3 Classification of 1958 couples based on race.

sum) are also shown in the second table. Observe that

$$\sum_{ij} p_{ij} = 1, \quad \sum_j p_{ij} = a_i, \quad \sum_i p_{ij} = b_j.$$

Let A denote the square matrix of values from the joint probability distribution. In the example under consideration the matrix is

$$\begin{bmatrix} 0.258 & 0.016 & 0.035 & 0.013 \\ 0.012 & 0.157 & 0.058 & 0.019 \\ 0.013 & 0.023 & 0.306 & 0.035 \\ 0.005 & 0.007 & 0.024 & 0.016 \end{bmatrix}.$$

Newman defined an *assortativity coefficient* as

$$a = \frac{tr(A) - sum(A^2)}{1 - sum(A^2)},$$

where $tr(A)$ is the sum of the diagonal elements and $sum(A^2)$ is the sum of all the entries of matrix A^2 (Newman, 2003a). Values of a are between -1 and 1. Positive values of a indicate *assortative* mixing. Negative values of a indicate *disassortative mixing*. When a is 0 the network is *non-assortative*. The aforementioned network has $a = 0.621$ indicating high assortative mixing.

The joint probability distribution table may be viewed as a complete bipartite graph with a weight on each edge. The weight represents the probability p_{ij} of edge $v_i v_j$. This analysis may be conducted for any weighted complete bipartite graph with the same number of vertices in both classes. Each weight must be divided by the total of all the weights in order to convert the weights into probabilities. If a weighted bipartite graph is not complete, it can be turned into a complete bipartite graph by adding the missing edges and assigning them a weight of zero.

A second approach to assortativity, applicable to an unweighted graph that is not necessarily bipartite, may be defined using vertex degrees. The basic idea is that some people are popular because they are part of a close-knit large group of people all of whom know each other. A graph shows assortative *mixing* if vertices that have high degree are joined by an edge to other vertices with high degree. If vertices with high degree are more likely to be joined to vertices with low degree, the graph is called disassortative. If there is no discernible relationship, the graph is called non-assortative (Newman, 2002). For digraphs similar definitions can be made separately for in-degrees and out-degrees.

In Pastor-Satorras et al. (2001) the authors analyzed the "Internet graph" for assortativity. They modeled the Internet by considering each ISP as a vertex. This makes sense because within each ISP, information is routed using an internal routing algorithm that may be different from other ISPs. So each ISP can be considered an independent unit of the Internet called an autonomous system (AS). The ISPs communicate between themselves using a specific routing algorithm called

Border Gateway Protocol. On 8 November 1997, 1998, and 1999, the Internet consisted of 3112 AS and 5450 edges, 3834 AS and 6990 edges, and 5287 AS and 10 100 edges, respectively. The mean degrees for these graphs were 3.5, 3.6, and 3.8, respectively. The authors calculated for each degree k, the mean degree of all the neighbors of the vertices of degree k. This gave a unique number for each degree k, and the resulting *assortativity function* can be plotted with degrees on the x-axis and the mean degree of neighbors of a specified degree on the y-axis. If the function is increasing, the graph is assortative, if the function is decreasing the graph is disassortative, and if the function is constant the graph is non-assortative. Using this measure, they found the Internet graphs of 1997, 1998, and 1999 to be disassortative.

A similar analysis can be conducted for the Erdős-1 collaboration graph shown in Figure 1.19. The assortativity function for this graph is shown in Figure 4.4. Observe, for example, that the value corresponding to vertex degree 20 is 19. This means that the average degree of all vertices in the graph adjacent to a vertex of degree 20 is 19. The function is increasing, indicating that this graph describes an assortative network.

A third approach to assortativity is to calculate the correlation coefficient of the degrees at either end of an edge (Newman, 2002). Given an edge $e = uv$, the remaining degrees of e at u and v are defined as

$$(deg(u) - 1, deg(v) - 1).$$

A joint probability distribution table with remaining degrees along the rows and columns can be prepared, where p_{ij} is the joint probability of remaining degrees x_i and x_j occurring for an edge. In this case the marginal probability a_i (and similarly b_j) is the probability that the remaining degree of a vertex at one end of an

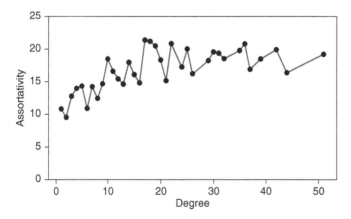

Figure 4.4 Assortativity function for the Erdős-1 collaboration graph.

edge is i. The covariance

$$\sigma_{XY} = \frac{1}{n}\sum_i (x_i - \mu_X)(y_i - \mu_Y) = \mu_{XY} - \mu_X\mu_Y$$

can be simplified as follows:

$$\sigma_{XY} = \sum_{i,j} x_i y_j p_{ij} - \sum_i x_i a_i \sum_j y_j b_j$$
$$= \sum_{i,j} x_i y_j p_{ij} - \sum_{i,j} x_i y_j a_i b_j$$
$$= \sum_{i,j} x_i y_j (p_{ij} - a_i b_j).$$

For a graph the marginal probabilities a_i and b_j are the same. Let σ be the standard deviation of remaining degree x_i occurring with probability a_i. The correlation coefficient is

$$r = \frac{\sigma_{XY}}{\sigma_X \sigma_Y} = \frac{1}{\sigma^2}\sum_{i,j} x_i x_j (p_{ij} - a_i a_j).$$

Newman computed the correlation coefficient of the following five social networks, two engineered networks, and three biological networks to determine their assortativity:

(1) Physics collaboration graph ($n = 52\ 909$ and $r = 0.363$) (Newman, 2001);
(2) Biology collaboration graph ($n = 1\ 520\ 251$ and $r = 0.127$) (Newman, 2001);
(3) Mathematics collaboration graph consisting of all mathematicians with Erdos number 1 and 2 ($n = 253\ 339$ and $r = 0.120$) (Grossman and Ion, 1995);
(4) Film actors collaboration graph ($n = 449\ 913$ and $r = 0.208$) (Watts and Strogatz, 1998);
(5) Company directors collaboration graph ($n = 7673$ and $r = 0.276$) (Davis et al., 2003);
(6) Internet graph ($n = 10\ 697$ and $r = -0.189$) (Chen et al., 2002);
(7) World-wide-web graph ($n = 269\ 504$ and $r = -0.065$) (Barabási and Albert, 1999);
(8) Protein-protein network ($n = 2115$ and $r = -0.156$) (Jeong et al., 2001);
(9) C. elegans connectome ($n = 307$ and $r = -0.163$) (Watts and Strogatz, 1998);
(10) Fresh-water food web ($n = 92$ and $r = -0.276$) (Martinez, 1991).

The five social networks are assortative, whereas the two engineered networks and three biological networks are disassortative. It has been confirmed experimentally that no matter what measure is used most social networks tend to be assortative, whereas biological networks and engineered networks tend to be disassortative.

We will end this section with a brief description of the network approach to systems biology taken from Aderem (2005). Systems biology is a comprehensive quantitative analysis of the manner in which all the components of a biological system interact functionally over time. The key concepts are emergence, robustness, and modularity. Emergence is the discovery of links among genes, proteins, and metabolites. Robustness means biological systems maintain their main functions even under external perturbations. Modularity is harder to explain and depends on context. A module in a network is a set of nodes that have strong interactions and a common function. Modularity contributes to robustness by confining damage to mostly separate groups of vertices. Modularity also decreases total failure of the system by containing the damage to one area. For a survey on complex networks in systems biology, see Costa et al. (2008).

4.4 Covert Networks

Covert networks are networks whose vertices and edges have an element of secrecy about them. Criminal networks such as those formed by insurgent groups, gangs, drug cartels, and arms traffickers are covert networks. It is difficult to identify the vertices and edges. The data is incomplete (who is in the network and who isn't) and the network is not static. According to Sparrow (1991), who gives the early history of network analysis in law enforcement, instead of looking at the presence or absence of a link, it may be better to look at the waxing and waning of links depending on time and the task at hand.[4] Not all covert social networks have negative connotations. Zegota, for instance, was a covert Roman Catholic underground organization that addressed the welfare needs of Jews in German-occupied Poland from 1942 to 1945. The Underground Railroad was a network of people who helped enslaved African Americans escape to the free states using secret routes and safe houses along the way. The newest covert network to enter our collective thinking is the COVID-19 network formed by contact tracing.

Covert networks have layers of interaction. For example, in a social network some of the people are linked professionally, some through personal friendships, some through social media, etc. Such networks are called multilayer networks. The layers may or may not have the same set of vertices. Criminal networks have layers that correspond to friendship ties, financial ties, religious ties, school ties, kinship ties, etc. Not all the members share the same sorts of ties and projecting the layers of networks into one large network hides valuable information.

---

4 See Sean Everton's book *Disrupting Dark Networks* Everton (2012) and Roberts and Everton (2011), which provide a description of data gathering methods and strategies for analyzing covert networks along with many references.

In epidemiology multiple pathogens coexist within the same host population interacting with each other while spreading in the population. Some compete with each other and some enhance or impair each other.

Roberts and Everton studied a multilayer network related to one particular criminal (2011). They calculated the mean degree, density, local mean clustering coefficient, and centralization indices for each of the layers. The number next to each network in the following text is its centralization index.

(1) Operational Network
  (a) *Communications Layer (0.412)*: Two vertices are adjacent if one sent a message to another.
  (b) *Logistics Layer (0.154)*: Two vertices are adjacent if they were both present in a key logistical location.
  (c) *Operations Layer (0.232)*: Two vertices are adjacent if both participated in the same operation.
  (d) *Financing Layer (0.073)*: Two vertices are adjacent if they work for the same business or foundation.
  (e) *Organizational Layer (0.326)*: Two vertices are adjacent if they are members of the same organization.
  (f) *Training Layer (0.157)*: Two vertices are adjacent if they participated in the same training camp.
(2) Trust Network
  (a) *Friendship Layer (0.129)*: Two vertices are adjacent if they are friends.
  (b) *Kinship Layer (0.022)*: Two vertices are adjacent if they are related.
  (c) *Religious Layer (0.049)*: Two vertices are adjacent if they are members of the same religious organization.
  (d) *School Layer (0.233)*: Two vertices are adjacent if they went to the same school.

Merging the layers gave one Operational Network and one Trust Network with centralization indices 0.448 and 0.216, respectively. The authors argued that merging layers hides the layers that exhibit the most centrality (the Communications Layer for the Operational Network and the School Layer for the Trust Network).

The authors also analyzed two additional networks: the Regional Schools Network and the Organizations Network (see Figure 4.5). In the Regional Schools Network the vertices are educational institutions. Two schools were linked if a member of the criminal network attended both. Three religious schools were found to be central. Therefore the authors recommended an early intervention strategy targeted toward these three schools that included infiltration and intelligence gathering. In the Organizations Network the vertices were criminal organizations and they were linked if they had at least one common member. The authors found that there was one obviously central organization (vertex *d*

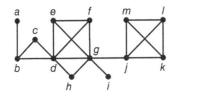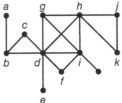

Figure 4.5 Regional Schools Network and Organizations Network.

in Figure 4.5). Its centrality was so significant that the authors recommended directing most efforts toward it.

Statistical analyses of observational studies should not be used to make broad predictions. Making predictions requires a very large amount of data and careful inferential methods. Prevailing techniques would require the surreptitious monitoring of a large number of people for decades raising numerous ethical questions. Even so, the results are unreliable as explained in subsequent paragraphs.

Credit card fraud detection methods are an example of good covert surveillance. With the permission of their customers, credit card companies gather every single credit card transaction and analyze them for patterns that indicate fraudulent activity. The set up allows for such large-scale automatic data gathering as part of the agreement in using a credit card. Careful documentation of every single financial transaction is a necessity for proper functioning of a civilized society. Doing the same with respect to human behavior is not only unethical, but also the results are unreliable. Human error in clandestine monitoring of other humans, or creating algorithms to do the same, is too great. Bias in judgment can never be removed due to the ends-justify-the-means environment and ample incentive to skew results. Moreover, when the government makes a mistake about an individual or group, the consequence for the government is negligible while the cost to the individual or group is tremendous; the cost differential is too large for an ethical democratic society. Besides, even a credit card company calls to confirm if a transaction is genuinely suspicious. If instead of such checks and balances, a credit card company relied solely on surreptitious monitoring, they would be out of business rather quickly when customers find a hold is put on their credit cards at the most inopportune times.

Earthquake detection methods are another good example of prediction. However, the algorithms that give good results for earthquakes are also applied to detect crime in neighborhoods (Shapiro, 2017). In mathematics we seek to find similarities in dissimilar objects, and situations and techniques that work well in one area often work well in another. There are limitations to this approach. If people committing small crimes are equivalent to little shocks before the big one, then using the same techniques makes perfect sense. But how effective is it to treat human behavior like an inorganic process? There is little cost to being wrong about a false-positive earthquake prediction compared with the cost of a false arrest borne by a human being.

We will end this section by mentioning two additional studies involving centrality measures in covert networks. In Baker and Faulkner (1993) the authors describe how an illegal price-fixing network was discovered using archival data including court records and sworn testimony. Centrality measures identified the important players in the network. They recommend using centrality measures, size of the network, and management level to determine whether a fine or a sentence should be imposed on a player. In Cunningham et al. (2013) the authors described how a grassroots organization, founded to protect Colombian peasants from harsh landowner policies, evolved into an international drug cartel. Centrality measures identified the key leaders tied to the transformation.

This chapter and the previous chapter, provide a set of techniques that can be used to analyze a large graph. The exercises in this section are well-suited for independent study and research projects, especially if a student constructs an original large graph, and better yet if an original large graph leads to an original technique like PageRank for webpages.

Exercises

4.1 For each of the five large graphs (karate club, Medici power network, dolphin social network, cat brain, and fly medulla) described in Exercise 3.5
   (a) Compare the graph with the corresponding random graph to determine if it is a small-world network;
   (b) Analyze the degree distribution to determine if the graph is a scale-free network; and
   (c) Determine assortativity using the methods described in Section 4.3.

Topics for Deeper Study

4.2 *Language*: How does a child learn the precise syntax of a language? An alternative to Noam Chomsky's prevailing "Principles and Parameters" theory is the Random Language model built on the assumption that sentences have a tree structure. Review this new model in DeGiuli (2019).

4.3 *Pattern Detection*: Suppose G is a graph with vertices representing email addresses and edges linking two addresses if more than a certain threshold number of emails are exchanged between them. Further suppose the graph is anonymized so the vertices are labeled by some unknown key, but the link structure is authentic. Such a situation might arise, for instance in large-scale gathering of packet data sent over the internet. What strategy would identify the induced subgraph of a predetermined set of vertices $x_1, x_2, \ldots, x_k$? One approach is to create in advance a set of email accounts $y_1, y_2, \ldots, y_p$ and link them to each other and to vertices $x_1, \ldots, x_k$ by sending emails in a very specific manner. Then this induced subgraph could be identified in the large graph along with the induced subgraph formed by $y_1, y_2, \ldots, y_p$. Review the approaches to this problem in the paper "Wherefore Art Thou R3579X" (Backstrom et al., 2007).

4.4 *Terrestrial Climate Networks*: A climate network is a complex network formed by linking locations on Earth using a measure of similarity based on climate data. Each vertex represents a location on Earth and has an associated time series of temperature, precipitation, pressure, etc. Vertices with highly correlated time series are joined by an edge. Climate networks are used to improve prediction of El Niño and La Niña events. Explore these types of networks in, for example, Tsonis et al. (2006) and Wiedermann et al. (2016).

4.5 *Temporal Networks*: A temporal network is a graph G whose edges are active at certain points in time. Associated to each edge e is a sequence of numbers $t_1, t_2, \ldots t_k$ representing the times that the edge e exists. Temporal networks are particulary useful in describing the spread of diseases. Explore these types of networks in, for example, Kempe et al. (2002) and Li et al. (2017).

5

Graph Algorithms

This chapter introduces graph algorithms, the third major topic in the book along with theory and applications. Graph algorithms is an interdisciplinary area covering both mathematics and computer science. A thorough understanding requires knowing computational complexity (see Appendix C), data structures (see Appendix D), and a programming language in order to implement the algorithms. Nonetheless, it is possible to understand the algorithms and their importance in a more theoretical manner.

Generally speaking graph algorithms can be divided into two broad classes: existence algorithms and optimization algorithms. An existence algorithm finds the property, invariant, or object of interest. An optimization algorithm finds the best option among a set of options. Section 5.1 presents two algorithms that determine whether or not a graph is connected; if the graph is connected, the algorithms find a spanning tree. They are examples of existence algorithms. Section 5.2 presents two algorithms for finding a minimum weight spanning tree in a weighted graph. Both use a "greedy strategy." They are examples of optimization algorithms. Section 5.3 presents Dijkstra's Shortest Path Algorithm for weighted graphs, another optimization algorithm.

Richard Karp introduced the concept of NP-complete in his 1972 paper "Reducibility among combinatorial problems" and highlighted several graph invariants as being intractable (Karp, 1972). As such it is worth studying the problems in this chapter since they have nice and efficient algorithms.

5.1 Traversal Algorithms

In this section we examine two algorithms to determine if a graph is connected: Depth-First Search (DFS) and Breadth-First Search (BFS). The input for both DFS and BFS is a graph and a starting vertex (call it s). Both algorithms search the graph by traversing edges from vertex to vertex, building a spanning tree as they proceed.

Graphs and Networks, First Edition. S. R. Kingan.

The strategy followed by DFS is to traverse deeper in the graph. Beginning with the starting vertex, the algorithm proceeds as far as possible before back-tracking. It terminates by producing the depth-first spanning tree if the graph is connected, or a statement that the graph is disconnected. On the other hand, BFS traverses the graph uniformly across the frontier of undiscovered vertices. It begins by visiting all the neighbors of the starting vertex s, after which it visits all the neighbors of the neighbors, and so on. If the graph is connected, it reaches every vertex from s and produces the breadth-first spanning tree. Moreover, the path from s to any vertex w in a breadth-first spanning tree is the path of shortest length from s to w. Therefore BFS also gives $d(s, w)$, for every vertex w distinct from s, and consequently the total distance $td(s)$.

A formal description of the DFS algorithm is presented in the following text. It has as inputs the graph G and a starting vertex s, as well as a set S of previously visited vertices and a set T of edges already in the spanning tree that is formed as the algorithm proceeds. The algorithm is recursive; that is, at some point the algorithm executes itself again with less of the graph yet to explore.

Algorithm 5.1.1. Depth-First Search

Input: A graph G, a set S of previously visited vertices (initially empty), a set T of edges (initially empty), and a starting vertex s.
Output: An indicator that is TRUE if G is connected and FALSE otherwise, and a list T of edges.

(1) Add the vertex s to S.
(2) **For each** $w \in N(s) - S$:
    Add the edge sw to T.
    Execute the algorithm with G, S, T, and w.
(3) **If** $S = V$
    Return TRUE
(4) **else**
    Return FALSE.

When the algorithm completes execution, if the result is TRUE, T will contain a spanning tree.

Let G be a graph on n vertices. The complexity[1] of the DFS algorithm is $O(n^2)$. To see this, observe that DFS visits each vertex exactly once and each edge at most twice, since each step involves traversing an edge; traversal is either forward or

---

1 The complexity of this algorithm, and all other algorithms discussed, is the worst-case computational complexity.

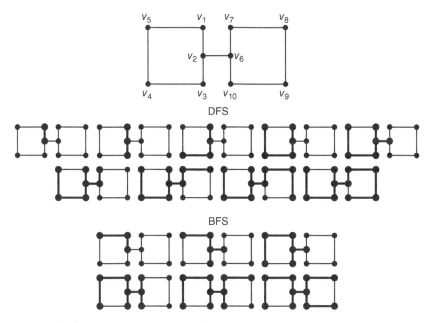

Figure 5.1 Step-by-step explanation of DFS and BFS.

backward. So the entire procedure investigates at most $2m$ edges. The number of steps is

$$n + 2m \leq n + 2 \binom{n}{2} = n + \frac{2n(n-1)}{2} = \frac{n + n^2 - n}{2} = n^2.$$

Consider the graph G shown in Figure 5.1. Assuming the starting vertex is v_1, DFS begins at v_1, and visits v_2, v_3, v_4, and v_5. It is unable to go further since all the neighbors of v_5 have been visited, therefore it back-tracks to the first available vertex with an unvisited edge, which is v_2, and then goes to v_6, v_7, v_8, v_9, v_{10}. At this stage all 10 vertices have been visited, so the algorithm stops and concludes that the graph is connected. If the algorithm backtracks and gets stuck at a vertex unable to go to any neighbor, but the number of visited vertices is less than n, it concludes the graph is disconnected.

The step-by-step execution of the DFS Algorithm with the starting vertex $s = v_1$ is presented in the following text:

- In the beginning $s = v_1$ and S and T are empty. Vertex v_1 is added to S. The neighbors of v_1 are v_2 and v_5. Vertex v_2, which has not been visited, is selected as a vertex adjacent to v_1 and edge $v_1 v_2$ is added to T. The algorithm is executed again with G, $S = \{v_1\}$, $T = \{v_1 v_2\}$, and $s = v_2$.

- During the second execution, v_2 is added to the list of visited vertices and v_3 is selected as a vertex adjacent to v_2 (v_1 cannot be selected, but v_6 could have been selected). Edge v_2v_3 is added to T. Then the algorithm is executed again with G, $S = \{v_1, v_2\}$, $T = \{v_1v_2, v_2v_3\}$, and $s = v_3$.
- During the third execution, v_3 is added to the list of visited vertices. The only non-visited vertex adjacent to v_3 is v_4, so it is selected for the next execution. Edge v_3v_4 is added to T and the algorithm is executed again with G, $S = \{v_1, v_2, v_3\}$, $T = \{v_1v_2, v_2v_3, v_3v_4\}$, and $s = v_4$.
- During the fourth execution, v_4 is added to the list of visited vertices. The only non-visited vertex adjacent to v_4 is v_5, so it is selected for the next execution. Edge v_4v_5 is added to T and the algorithm is executed again with G, $S = \{v_1, v_2, v_3, v_4\}$, $T = \{v_1v_2, v_2v_3, v_3v_4, v_4v_5\}$, and $s = v_5$.
- When the algorithm is executed with $s = v_5$ and $S = \{v_1, v_2, v_3, v_4\}$, there are no non-visited vertices adjacent to v_5. Therefore v_5 is added to the list of visited vertices and then this execution of the algorithm terminates, as do the fourth and third executions.
- When the second execution of the algorithm resumes, there is one more non-visited vertex adjacent to v_2, namely v_6. Edge v_2v_6 is added to T and the algorithm is executed again with G, $S = \{v_1, v_2, v_3, v_4, v_5\}$, $T = \{v_1v_2, v_2v_3, v_3v_4, v_4v_5, v_2v_6\}$, and $s = v_6$.
- The algorithm continues similarly, adding vertices v_7, v_8, v_9, and v_{10} to S and edges v_6v_7, v_7v_8, v_8v_9, and v_9v_{10} to T, respectively, in its next executions. When $s = v_{10}$ there are no further non-visited adjacent vertices to consider, and also no further non-visited adjacent vertices in any earlier executions, so the first execution of the algorithm finally terminates, with spanning tree

$$T = \{v_1v_2, v_2v_3, v_3v_4, v_4v_5, v_2v_6, v_6v_7, v_7v_8, v_8v_9, v_9v_{10}\}.$$

The DFS algorithm keeps track of the vertices it still needs to visit through the data associated with the "trail," or *stack*, of executions that have been put on hold during recursion. In the aforementioned example, by the time the algorithm is executed with $s = v_5$, the vertices waiting to be visited in previous executions are:

$$s = v_1, N(s) - S = \{v_2, v_5\}\,(v_5 \text{ still has to be visited})$$
$$s = v_2, N(s) - S = \{v_3, v_6\}\,(v_6 \text{ still has to be visited})$$
$$s = v_3, N(s) - S = \{v_4\}$$
$$s = v_4, N(s) - S = \{v_5\}.$$

The BFS algorithm keeps track of vertices yet to be visited differently, using a mechanism called a *queue*. Stacks and queues are defined in Appendix D. Here we will simply assume that a queue is a list of items. Items can be added to the queue and removed from the queue when it is not empty, and items are removed in the same order they were added.

Algorithm 5.1.2. Breadth-First Search

Input: A graph G and a starting vertex s.
Output: An indicator that is TRUE if G is connected and FALSE otherwise, and a list T of edges.

(1) Let T be an empty set of edges.
(2) Let S be a set of vertices initially containing only s.
(3) Let Q be an empty queue. Add s to Q.
(4) **While** Q is not empty:
      Remove a vertex v from Q.
      For each $w \in N(v) - S$:
      Add vertex w to S.
      Add edge vw to T.
      Add vertex w to Q.
(5) **If** $S = V$
      Return TRUE
(6) **else**
      Return FALSE.

As in the case of DFS, when the BFS algorithm completes execution, if the result is TRUE, T will contain a spanning tree. Since BFS visits each vertex and edge exactly once, the number of steps is

$$n + m \leq n + \frac{n(n-1)}{2} = \frac{2n + n^2 - n}{2} = \frac{n^2 + n}{2}.$$

Therefore the computational complexity of BFS is also $O(n^2)$.

Consider again the graph G shown in Figure 5.1. Assuming the starting vertex is v_1, BFS begins at v_1 and visits neighbors v_2 and v_5. Moving to v_2, BFS visits v_3 and v_6, and then moving to v_5 it visits v_4. From v_6 it visits v_7 and v_{10}. From v_7 it visits v_8 and from v_{10} it visits v_9. Since it traversed all 10 vertices, BFS stops and concludes the graph is connected. The step-by-step execution of the algorithm with starting vertex $s = v_1$ is presented in the following text:

- At the beginning of the first iteration of the loop at line 4, $S = \{v_1\}$, $Q = [v_1]$, and $T = \phi$. The value removed from Q is $v = v_1$. Neither of the neighbors of v_1, v_2, and v_5 are in S, so both are added to S, both are added to Q, and edges v_1v_2 and v_1v_5 are added to T.
- At the second iteration, $S = \{v_1, v_2, v_5\}$, $Q = [v_2, v_5]$, and $T = \{v_1v_2, v_1v_5\}$. The value removed from Q is $v = v_2$, which has neighbors v_1, v_3, and v_6. Since neither v_3 nor v_6 is in S, both are added to S, both are added to Q and edges v_2v_3 and v_2v_6 are added to T.

- At the third iteration, $S = \{v_1, v_2, v_5, v_3, v_6\}$, $Q = [v_5, v_3, v_6]$, and $T = \{v_1 v_2, v_1 v_5, v_2 v_3, v_2 v_6\}$. The value removed from Q is $v = v_5$, which has neighbors v_1 and v_4. Since $v_4 \notin S$, v_4 is added to S, v_4 is added to Q, and edge $v_5 v_4$ is added to T.

- At the fourth iteration, $S = \{v_1, v_2, v_5, v_3, v_6, v_4\}$, $Q = [v_3, v_6, v_4]$, and $T = \{v_1 v_2, v_1 v_5, v_2 v_3, v_2 v_6, v_5 v_4\}$. The value removed from Q is $v = v_3$, which has no neighbors not in S.

- At the fifth iteration, the value removed from Q is $v = v_6$, which has neighbors v_7 and v_{10}, neither of which are in S. Both are added to S and to Q, and edges $v_6 v_7$ and $v_6 v_{10}$ are added to T.

- At the sixth iteration, $S = \{v_1, v_2, v_5, v_3, v_6, v_4, v_7, v_{10}\}$, $Q = [v_4, v_7, v_{10}]$, and $T = \{v_1 v_2, v_1 v_5, v_2 v_3, v_2 v_6, v_5 v_4, v_6 v_7, v_6 v_{10}\}$. The value removed from Q is $v = v_4$, which has no neighbors not already in S.

- At the seventh iteration, the value removed from Q is $v = v_7$, which has neighbors v_6 and v_8. Since $v_8 \notin S$, it is added to S and to Q, and edge $v_7 v_8$ is added to T.

- At the eighth iteration, $S = \{v_1, v_2, v_5, v_3, v_6, v_4, v_7, v_{10}, v_8\}$, $Q = [v_{10}, v_8]$, and $T = \{v_1 v_2, v_1 v_5, v_2 v_3, v_2 v_6, v_5 v_4, v_6 v_7, v_6 v_{10}, v_7 v_8\}$. The value removed from Q is $v = v_{10}$, which has neighbors v_6 and v_9. Since $v_9 \notin S$, v_9 is added to S and Q, and the edge $v_{10} v_9$ is added to T.

- At the ninth iteration, $S = \{v_1, v_2, v_5, v_3, v_6, v_4, v_7, v_{10}, v_8, v_9\}$, $Q = [v_8, v_9]$, and $T = \{v_1 v_2, v_1 v_5, v_2 v_3, v_2 v_6, v_5 v_4, v_6 v_7, v_6 v_{10}, v_7 v_8, v_{10} v_9\}$. The value removed from Q is $v = v_8$, which has no neighbors not in S.

- At the tenth iteration, the value removed from Q is $v = v_9$, which also has no neighbors not in S. At this point Q is empty, so the algorithm terminates with spanning tree

$$T = \{v_1 v_2, v_1 v_5, v_2 v_3, v_2 v_6, v_5 v_4, v_6 v_7, v_6 v_{10}, v_7 v_8, v_{10} v_9\}.$$

Observe that the DFS spanning tree may be different from the BFS spanning tree.

For a high level understanding this much is enough. Appendix D has details that provide insight into how these two navigation methods originated and the difference between them.

5.2 Greedy Algorithms

Finding a spanning tree in a graph is an existence problem, whereas finding a minimum weight spanning tree in a weighted graph is an optimization problem.

Let G be a connected weighted graph. We may think of the weights as a function w that assigns a real number to each edge. If $e = uv$ is an edge, denote the weight of e as $w(e)$ or $w(uv)$.

The weight of a spanning tree T is the sum of the weights of the edges in T. In other words

$$w(T) = \sum_{e \in T} w(e).$$

A minimum weight spanning tree has lowest weight among all spanning trees. In short it is called a *minimum spanning tree*. Note that there may be more than one minimum spanning tree in a graph. The problem of finding a minimum spanning tree is called the *Minimum Spanning Tree Problem.*

Following Cormen et al. (2001), we will describe a "greedy strategy" to find a minimum spanning tree. Kruskal's Algorithm and Prim's Algorithm both use this greedy strategy, but in different ways. A greedy algorithm always makes the choice that looks best at the moment. It makes a locally optimal choice in the hope that this choice will lead to a globally optimal solution. The greedy strategy is quite simple, a minimum spanning tree T is grown from an initially empty set of edges A by repeatedly adding a minimum weight edge e to A in such a way that $A \cup e$ remains acyclic. This edge e is called a "safe edge". The key questions are do safe exist and if they do, then how should we select them? Establishing the existence of safe edges requires a proof. How to select safe edge requires an algorithm.

Algorithm 5.2.1. Generic Minimum Spanning Tree Algorithm

Input: A connected graph G and a positive weight function on the edges.
Output: A set of edges that forms a minimum spanning tree.

(1) Let A be an empty set of edges.
(2) **While** A does not form a spanning tree:
      Find an edge e of minimum weight that is safe for A
      Add e to A.
(3) **Return** A.

Before proving that safe edges exist we need a lemma on trees and an additional piece of terminology. Let G be a graph and X be a set of edges of G. We will denote X with an additional edge e as $X \cup e$ instead of $X \cup \{e\}$ and X with edge e removed as $X - e$ instead of $X - \{e\}$.

Lemma 5.2.2. *Let G be a connected graph and T_1 and T_2 be two spanning trees. If edge $e_1 \in T_1 - T_2$, then there exists an edge $e_2 \in T_2 - T_1$ such that $(T_1 - e_1) \cup e_2$ is a spanning tree.*

Proof. First observe that since $e_1 \in T_1$, $T_1 - e_1$ is disconnected and has two connected components, say X and Y, where one end vertex of e_1 is in X and the other in Y. Second observe that $|T_1| = |T_2|$ since both are spanning trees of G. Since $e_1 \notin T_2$, and every vertex of G is in X or Y, there exists an edge $e_2 \in T_2 - T_1$ with one end vertex in X and the other in Y. Clearly $(T_1 - e_1) \cup e_2$ is connected. Moreover, $(T_1 - e_1) \cup e_2$ is acyclic because if it had a cycle, then the cycle must contain e_2, which is not possible since e_2 is a bridge in $(T_1 - e_1) \cup e_2$. So $(T_1 - e_1) \cup e_2$ is a tree. Finally, since $(T_1 - e_1) \cup e_2$ has the same number of edges as T_1, it is a spanning tree. □

Let $G = (V, E)$ be a weighted graph and $S \subset V$. An edge e *crosses* the partition $(S, V - S)$ if one of its end vertices is in S and the other is in $V - S$. Let $A \subset T$, where T is a minimum spanning tree. An edge e is *safe* for A if $A \cup e$ is also a subset of a minimum spanning tree, not necessarily T.

Theorem 5.2.3. *Let $G = (V, E)$ be a weighed graph and T be a minimum spanning tree. Let $A \subset T$ and let $(S, V - S)$ be a partition of V such that no edge in A crosses $(S, V - S)$. Let e be a minimum weight edge that crosses $(S, V - S)$. Then e is a safe edge for A.*

Proof. Observe that $e \notin A$ since e crosses $(S, V - S)$ and no edge of A crosses $(S, V - S)$. If $e \in T$, then clearly e is a safe edge since $A \cup e$ is a subset of a minimum spanning tree, namely T. Therefore suppose $e \notin T$. We will construct another minimum spanning tree T' such that $A \cup e \subseteq T'$, thereby showing e is safe for A. Let u and v be the end vertices of edge e. Since e crosses $(S, V - S)$ we may assume $u \in S$ and $v \in V - S$. However, since $e \notin T$, e forms a cycle with the path P along T joining u and v. Moreover, since e crosses $(S, V - S)$, there is at least one edge e' in T along path P that also crosses $(S, V - S)$. Observe that $e' \notin A$ since no edge in A crosses $(S, V - S)$.

Consider $T' = (T - e') \cup e$. By the same explanation as Lemma 5.2.2, T' is a spanning tree. We will prove that T' is a minimum spanning tree. First note that since T is a minimum spanning tree $w(T') \geq w(T)$. Since e is an edge of minimum weight that crosses $(S, V - S)$ and e' also crosses $(S, V - S)$, $w(e') \geq w(e)$. Therefore

$$w(T') = w(T) - w(e') + w(e) \leq w(T)$$

and we may conclude that

$$w(T) = w(T').$$

So T' is a minimum spanning tree. Lastly, since $e \notin A$, $A \cup e \subseteq T'$. Therefore e is a safe edge for A. □

In 1956 Joseph Kruskal described one way of selecting safe edges (Kruskal, 1956). The next year Robert Prim described another way (Prim, 1957). In the case of Kruskal's algorithm, a forest is grown by starting with the lowest weight edge and at each stage the safe edge selected has lowest weight among available edges anywhere in the graph. In the case of Prim's algorithm, a tree is grown by starting with a vertex and at each stage the safe edge selected has lowest weight among the neighbors of already visited vertices.

Algorithm 5.2.4. Kruskal's Algorithm

Input: A connected graph $G = (V, E)$ and a weight function on the edges.
Output: A set of edges that forms a minimum spanning tree.

(1) Let A be an empty set of edges.
(2) Let m be $|E|$.
(3) Arrange the edges of E into a list in non-decreasing order by weight.
(4) **For each** $i = 1, \dots, m$:
  If the edge e at position i in the list connects two previously disconnected components of $G' = (V, A)$:
  Add e to A.
(5) **Return** A.

Consider an edge e added at iteration i of Step 4 in Kruskal's Algorithm. Let C denote one of the disconnected components of $G' = (V, A)$ connected by e, and let S be the set of vertices in V incident with some edge in C. Then no edge in A crosses $(S, V - S)$, e does cross $(S, V - S)$, and because the edges are considered in non-decreasing order by weight, it is a minimum weight edge that does so. Therefore by Theorem 5.2.3, e is a safe edge for A.

The complexity of Kruskal's Algorithm is $O(m \log n)$, where m is the number of edges and n is the number of vertices in the input graph. While the algorithm's main loop is $O(m)$, sorting the edges in non-descending order is an $O(m \log n)$ operation ((Cormen et al., 2001), p. 570).

Figure 5.2 shows a step-by-step implementation of Kruskal's algorithm.

- At Step 3, all the edges are sorted in increasing order by weight:

$$[v_3 v_6, v_2 v_6, v_4 v_5, v_1 v_2, v_2 v_3, v_1 v_6, v_5 v_6, v_3 v_5, v_3 v_4].$$

- At this point $T = \phi$ so the connected components of $G' = (V, T)$ are just the individual vertices: $\{v_1\}$, $\{v_2\}$, $\{v_3\}$, $\{v_4\}$, $\{v_5\}$, and $\{v_6\}$.
- At the first iteration of the loop at Step 4, since edge $v_3 v_6$ connects components $\{v_3\}$ and $\{v_6\}$, it is added to T. Now $T = \{v_3 v_6\}$.

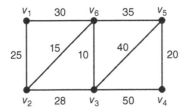

Kruskal's algorithm

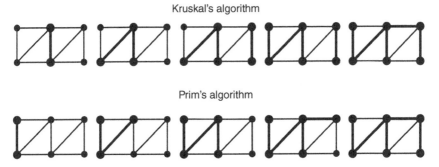

Prim's algorithm

Figure 5.2 Kruskal's and Prim's algorithms.

- At the second iteration, the connected components of G' are $\{v_1\}$, $\{v_2\}$, $\{v_3, v_6\}$, $\{v_4\}$, and $\{v_5\}$. Edge $v_2 v_6$ connects components $\{v_2\}$ and $\{v_3, v_6\}$, so $v_2 v_6$ is added to T. Now $T = \{v_2 v_6, v_3 v_6\}$.
- At the third iteration, the connected components of G' are $\{v_1\}$, $\{v_2, v_3, v_6\}$, $\{v_4\}$, and $\{v_5\}$. Edge $v_4 v_5$ connects components $\{v_4\}$ and $\{v_5\}$, so it is added to T. Now $T = \{v_2 v_6, v_3 v_6, v_4 v_5\}$.
- At the fourth iteration, the connected components of G' are $\{v_1\}$, $\{v_2, v_3, v_6\}$, and $\{v_4, v_5\}$. Edge $v_1 v_2$ connects components $\{v_1\}$ and $\{v_2, v_3, v_6\}$, so it is added to T. Now $T = \{v_1 v_2, v_2 v_6, v_3 v_6, v_4 v_5\}$.
- At the fifth iteration, the connected components of G' are $\{v_1, v_2, v_3, v_6\}$ and $\{v_4, v_5\}$. Edge $v_2 v_3$ does not connect any components, so it is not added to T.
- At the sixth iteration, $v_1 v_6$ also does not connect any components of G', so it is not added to T.
- At the seventh iteration, $v_5 v_6$ connects components $\{v_1, v_2, v_3, v_6\}$ and $\{v_4, v_5\}$, so it is added to T. Now $T = \{v_1 v_2, v_2 v_6, v_3 v_6, v_4 v_5, v_5 v_6\}$.
- At the eighth and ninth iterations G' has only one component, so edges $v_3 v_5$ and $v_3 v_4$ are not added to T.

When the algorithm terminates, the minimum spaning tree is

$$T = \{v_1 v_2, v_2 v_6, v_3 v_6, v_4 v_5, v_5 v_6\}.$$

Algorithm 5.2.5. Prim's Algorithm

Input: A connected graph $G = (V, E)$ and a positive weight function w on the edges.

Output: A set of edges that form a minimum spanning tree.

(1) Choose a starting vertex u at random.
(2) Let S be a set of vertices initially containing only u.
(3) Let A be an empty set of edges.
(4) **While** $|S| < |V|$:

Choose an edge $e = uv$ in E such that:
(a) One of u and v is in S and the other is not. Say $u \in S$ and $v \notin S$.
(b) For any other edge e' with only one vertex in S, $w(e) \le w(e')$.
Add v to S.
Add e to A.

(5) **Return** A.

At each stage of Prim's Algorithm, no edge in A crosses $(S, V - S)$. The edge e that is added does cross $(S, V - S)$, and is selected to be the minimum weight edge that does so. Therefore by Theorem 5.2.3, e is a safe edge for A.

The complexity of Prim's Algorithm is $O(m + n \log n)$, where m is the number of edges and n is the number of vertices ((Cormen et al., 2001), p. 573).

The step-by-step details of the application of Prim's Algorithm to the graph in Figure 5.2 are as follows:

- At Step 1, the algorithm chooses a starting vertex at random. Let us assume that vertex v_1 is chosen. After Steps 2 and 3, $S = \{v_1\}$ and $T = \phi$.
- At the first iteration of the loop at Step 4, edges $v_1 v_6$, and $v_1 v_2$ each have one end in S and one end not in S. Since $v_1 v_2$ has lower weight than $v_1 v_6$, it is added to T and v_2 is added to S.
- At the second iteration, $S = \{v_1, v_2\}$ and $T = \{v_1 v_2\}$. Edges $v_1 v_6$, $v_2 v_6$ and $v_2 v_3$ have one end in S and one end outside of S. Of these, $v_2 v_6$ has the lowest weight, so $v_2 v_6$ is added to T and v_6 is added to S.
- At the third iteration, $S = \{v_1, v_2, v_6\}$ and $T = \{v_1 v_2, v_2 v_6\}$. Edges $v_2 v_3$, $v_5 v_6$, and $v_3 v_6$ have one end in S and one end outside of S. Of these, $v_3 v_6$ has the lowest weight, so $v_3 v_6$ is added to T and v_3 is added to S.
- At the fourth iteration, $S = \{v_1, v_2, v_3, v_6\}$, and $T = \{v_1 v_2, v_2 v_6, v_3 v_6\}$. Edges $v_5 v_6$, $v_3 v_5$ and $v_3 v_4$ all have one end in S and one end outside of S. Of these, $v_5 v_6$ has the lowest weight, so it is added to T and v_5 is added to S.
- At the fifth iteration, $S = \{v_1, v_2, v_3, v_5, v_6\}$ and $T = \{v_1 v_2, v_2 v_6, v_3 v_6, v_5 v_6\}$. Edges $v_4 v_5$ and $v_3 v_4$ have one end in S and one end outside of S. Of these, $v_4 v_5$ has the lower weight, so $v_4 v_5$ is added to T and v_4 is added to S. After this iteration $S = V$ so the loop terminates.

When the algorithm is done, $T = \{v_1 v_2, v_2 v_6, v_3 v_6, v_4 v_5, v_5 v_6\}$ is a minimum spanning tree.

Note that in this example, the minimum spanning tree is the same using both Prim's Algorithm and Kruskal's Algorithm. This is not necessarily always the case when edges have the same weights.

5.3 Shortest Path Algorithms

Computer scientist Edsger Dijkstra gave the first shortest path algorithm for a weighted directed graph (Dijkstra, 1959). Let G be a directed graph with a non-negative weight function w, where $w(u, v) = \infty$ if $(u, v) \notin E(G)$. Let P be a directed path from s to t given by

$$s = v_0, v_1, v_2, \ldots, v_k = t,$$

where (v_i, v_{i+1}) is an arc for $0 \leq i \leq k - 1$. The *weight* of P, denoted by $w(P)$, is defined as

$$w(P) = w(v_0, v_1) + w(v_1, v_2) + \cdots + w(v_{k-1}, v_k).$$

The *shortest path weight* from s to t, denoted by $\delta(s, t)$, is the minimum value of $w(P)$, where the minimum is taken over every directed path P from s to t, assuming at least one such path exists. A *shortest path* from s to t in a weighed graph is defined as any path P with $w(P) = \delta(s, t)$. For example, in the weighted directed graph shown in Figure 5.3, $s v_2 v_6 v_8 t$ is the shortest $s - t$ path and it has four edges and weight 11. Contrast this with $s v_2 v_5 t$, which has three edges, but is not the shortest $s - t$ path because its weight is 15.

Finding the shortest path between two vertices in a weighted graph (called the *Shortest Path Problem*) comes up in many situations. In operations research problems, vertices correspond to stages in a manufacturing process and the weight on a path from s to t corresponds to the cost of going from stage s to stage t. The goal is to find a sequence of activities from beginning to end that minimizes the cost. Map websites give driving directions based on shortest path algorithms. The exact

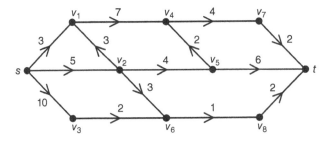

Figure 5.3 A Weighted Digraph.

algorithms used are proprietary and take into consideration various different definitions of "weight" including not only distance, but also travel time in traffic, road quality, and tolls. The core, however, is a shortest path algorithm.

In this section we will assume the weights are non-negative. A key idea used in Dijkstra's Algorithm is the optimal substructure property of shortest paths (Lemma 2.2.3). A shortest path from s to t contains within it a shortest path from s to any other vertex in the path before t.

Second, Dijkstra's Algorithm is based on BFS for digraphs, which is like Algorithm 5.1.2, except that edges are only followed along their direction. Dijkstra's Algorithm advances across the frontier of undiscovered vertices, but it also has to keep track of weights.

We need one more piece of notation before presenting the algorithm. We will use $d(v)$, where v is a vertex in the graph, to represent the length of the shortest path from s to v known at each step. The algorithm starts with $d(s) = 0$ and $d(v) = \infty$ for all vertices $v \neq s$, and assigns values to $d(v)$ as it proceeds and discovers paths of finite length. Dijkstra's algorithm presented in the following text is taken from Roberts (2005).

Algorithm 5.3.1. Dijkstra's Algorithm

Input: A directed graph G with a non-negative weight function w on pairs of vertices, and two vertices s and t.
   Output: A shortest path from s to t.

(1) Let A be a set of vertices initially containing only s.
(2) Let B be an empty set of arcs.
(3) Set $d(s)$ to 0 and $d(v)$ to ∞ for all $v \neq s$.
(4) **While** $t \notin A$:
   For each pair of vertices $u \in A$, $v \notin A$:
      Let $\alpha(u, v)$ be equal to $d(u) + w(u, v)$.
   If no finite values have been assigned for $\alpha(u, v)$:
      STOP. There are no paths from s to t in G.
   else
      Let $u, v \in V(G)$ such that $\alpha(u, v)$ is minimal. (Note that $\alpha(u, v)$ need
         not have been added in the most recent iteration.)
      Add v to A.
      Add (u, v) to B.
      Let $d(v)$ be $\alpha(u, v)$.

When the algorithm terminates, if $t \in A$, then B will contain a shortest path from s to t.

The step-by-step details of the application of Algorithm 5.3.1 to the graph in Figure 5.4 to find a shortest path from s to t are as follows:

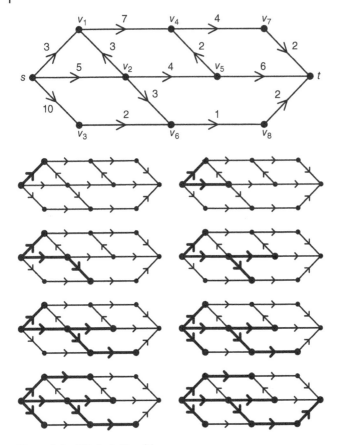

Figure 5.4 Dijkstra's Algorithm.

- At the beginning of the algorithm, $A = \{s\}$ and $B = \phi$. In the first iteration of the loop at Step 4, we compute the following finite values:

$$a(s, v_1) = d(s) + w(s, v_1) = 0 + 3 = 3$$

$$a(s, v_2) = d(s) + w(s, v_2) = 0 + 5 = 5$$

$$a(s, v_3) = d(s) + w(s, v_3) = 0 + 10 = 10.$$

All other values are infinite, because there are no arcs from s to any other vertices. The smallest a value is 3, so v_1 is added to A, (s, v_1) is added to B, and $d(v_1)$ is set to 3. Now $A = \{s, v_1\}$ and $B = \{(s, v_1)\}$. For all vertices v other than s and v_1, $d(v)$ remains ∞.

- In the second iteration we compute the following additional finite α value:

$$\alpha(v_1, v_4) = d(v_1) + w(v_1, v_4) = 3 + 7 = 10.$$

Again, other than this value and the values for $\alpha(s, v_1)$, $\alpha(s, v_2)$, and $\alpha(s, v_3)$ earlier, all other α values are infinite. Since the smallest value is now $5 = \alpha(s, v_2)$, v_2 is added to A, (s, v_2) is added to B, and $d(v_2)$ is set to 5. Now $A = \{s, v_1, v_2\}$ and $B = \{(s, v_1), (s, v_2)\}$.

- In the third iteration we compute the following additional finite α values:

$$\alpha(v_2, v_5) = d(v_2) + w(v_2, v_5) = 5 + 4 = 9$$
$$\alpha(v_2, v_6) = d(v_2) + w(v_2, v_6) = 5 + 3 = 8.$$

Since the minimum α value is now $8 = \alpha(v_2, v_6)$, v_6 is added to A, (v_2, v_6) is added to B, and $d(v_6)$ is set to 8. Now $A = \{s, v_1, v_2, v_6\}$ and $B = \{(s, v_1), (s, v_2), (v_2, v_6)\}$.

- In the fourth iteration we compute only one more finite α value:

$$\alpha(v_6, v_8) = d(v_6) + w(v_6, v_8) = 8 + 1 = 9.$$

Now both $\alpha(v_2, v_5)$ and $\alpha(v_6, v_8)$ are 9, so we may choose either. Choosing v_5, we add v_5 to A, add (v_2, v_5) to B, and set $d(v_5)$ to 9. Now $A = \{s, v_1, v_2, v_6, v_5\}$ and $B = \{(s, v_1), (s, v_2), (v_2, v_6), (v_2, v_5)\}$.

- In the fifth iteration we compute two more finite α values:

$$\alpha(v_5, v_4) = d(v_5) + w(v_5, v_4) = 9 + 2 = 11$$
$$\alpha(v_5, t) = d(v_5) + w(v_5, t) = 9 + 6 = 15.$$

Since the minimum α value is now $9 = \alpha(v_6, v_8)$, we add v_8 to A, add (v_6, v_8) to B, and set $d(v_8) = 9$. Even though we have now considered a path to t, it is not selected because another vertex has a smaller α value. Now $A = \{s, v_1, v_2, v_6, v_5, v_8\}$ and $B = \{(s, v_1), (s, v_2), (v_2, v_6), (v_2, v_5), (v_6, v_8)\}$.

- In the sixth iteration we compute one more finite α value:

$$\alpha(v_8, t) = d(v_8) + w(v_8, t) = 9 + 2 = 11.$$

The minimum α values are now $10 = \alpha(s, v_3) = \alpha(v_1, v_4)$, so we may choose either. Choosing v_3, v_3 is added to A, (s, v_3) is added to B, and $d(v_3)$ is set to 10. Now $A = \{s, v_1, v_2, v_6, v_5, v_8, v_3\}$ and $B = \{(s, v_1), (s, v_2), (v_2, v_6), (v_2, v_5), (v_6, v_8), (s, v_3)\}$.

- In the seventh iteration no new finite α values are computed, because the only vertex connected via an arc from v_3 is v_6 and $v_6 \in A$ already. Since $10 = \alpha(v_1, v_4)$ is still the minimal value, v_4 is added to A, (v_1, v_4) is added to B, and $d(v_4)$ is set to 10. Now $A = \{s, v_1, v_2, v_6, v_5, v_8, v_3, v_4\}$ and $B = \{(s, v_1), (s, v_2), (v_2, v_6), (v_2, v_5), (v_6, v_8), (s, v_3), (v_1, v_4)\}$.

- In the eighth iteration we compute one new finite α value:

$$\alpha(v_4, v_7) = d(v_4) + w(v_4, v_7) = 10 + 4 = 14.$$

The minimum value of α is now $11 = \alpha(v_8, t)$, so t is added to A, (v_8, t) is added to B and $d(t)$ is set to 11. Now $A = \{s, v_1, v_2, v_6, v_5, v_8, v_3, v_4, t\}$ and $B = \{(s, v_1), (s, v_2), (v_2, v_6), (v_2, v_5), (v_6, v_8), (s, v_3), (v_1, v_4), (v_8, t)\}$. Since $t \in A$ the algorithm terminates.

When the algorithm terminates, we can find the shortest path $s v_2 v_6 v_8 t$ in B, shown in bold:

$$B = \{(s, v_1), \mathbf{(s, v_2), (v_2, v_6)}, (v_2, v_5), \mathbf{(v_6, v_8)}, (s, v_3), (v_1, v_4), \mathbf{(v_8, t)}\}.$$

Algorithm 5.3.1 determines the shortest path in a directed and weighted graph from a specified vertex s to another specified vertex t, if it exists. It is what is called a "single-pair shortest-path algorithm." The algorithm can easily be modified to determine the shortest path from s to *every* other vertex t, and thus can also be referred to as a "single-source shortest-path algorithm," where source refers to the starting vertex. Specifically, if G is connected, simply change the condition in Step 4 of the algorithm from $t \notin A$ to $A \neq V(G)$. The complexity of Dijkstra's Algorithm is $O(n \log n + m)$ ((Cormen et al., 2001), p. 599).

Exercises

5.1 Consider the graph in Figure 3.3.
  (a) Apply Algorithms 5.1.1 and 5.1.2 to obtain the DFS spanning tree and BFS spanning tree starting with vertex a.
  (b) Use Algorithm 5.1.2 to find the eccentricity and the total distance of each vertex.

5.2 Consider the strongly connected digraph in Figure 3.3. Adjust Algorithm 5.1.2 to follow directed edges and use it to find the eccentricity and the total distance of each vertex.

5.3 Consider the weighted graphs in Figure 5.5. Apply Algorithms 5.2.4 and 5.2.5 to find minimum spanning trees.

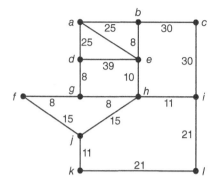 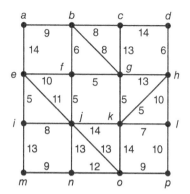

Figure 5.5 Examples of weighted graphs.

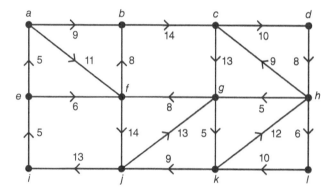

Figure 5.6 Example of a weighted digraph.

5.4 Consider the weighted directed graph in Figure 5.6. Apply Algorithm 5.3.1 to find the shortest path between every pair of vertices.

5.5 Prove that the set of cycles of a graph satisfy the following properties:
(i) If C_1 and C_2 are cycles such that $C_1 \subseteq C_2$, then $C_1 = C_2$; and
(ii) If C_1 and C_2 are cycles and $e \in C_1 \cap C_2$, then there is a cycle C_3 such that $C_3 \subseteq (C_1 \cup C_2) - e$.

5.6 Prove that sets of acyclic edges of a graph satisfy the following properties:
(i) If I_1 is a set of acyclic edges and $I_2 \subseteq I_1$, then I_2 is a set of acyclic edges; and
(ii) If I_1 and I_2 are two sets of acyclic edges such that $|I_1| < |I_2|$, then there is an edge $e \in I_2 - I_1$ such that $I_1 \cup e$ is a set of acyclic edges.

Topics for Deeper Study

5.7 Matroids have many cryptomorphic definitions. Prove that the following definitions of a matroid are equivalent.

(a) A matroid $(E, \mathcal{B})$ is a set of elements E together with a family $\mathcal{B}$ of subsets of E called bases[2] that satisfies:

(B1) $\mathcal{B} \neq \phi$; and

(B2) If $B_1, B_2 \in \mathcal{B}$ and $x \in B_1 - B_2$, then there exists $y \in B_2 - B_1$ such that $(B_1 - x) \cup y \in \mathcal{B}$.

(b) A matroid $(E, \mathcal{C})$ is a set of elements E together with a family $\mathcal{C}$ of subsets of E called circuits that satisfies:

(C1) $\phi \notin \mathcal{C}$;

(C2) If $C_1, C_2 \in \mathcal{C}$ such that $C_1 \subseteq C_2$, then $C_1 = C_2$; and

(C3) If $C_1, C_2 \in \mathcal{C}$ and $e \in C_1 \cap C_2$, then there exists $C_3 \in \mathcal{C}$ such that $C_3 \subseteq (C_1 \cup C_2) - e$.

(c) A matroid $(E, \mathcal{I})$ is a set of elements E together with a family $\mathcal{I}$ of subsets of E called independent sets that satisfies:

(I1) $\phi \in \mathcal{I}$;

(I2) If $I_1 \in \mathcal{I}$ and $I_2 \subseteq I_1$, then $I_2 \in \mathcal{I}$; and

(I3) If $I_1, I_2 \in \mathcal{I}$ such that $|I_1| < |I_2|$, then there exists $e \in I_2 - I_1$ such that $I_1 \cup e \in \mathcal{I}$.

5.8 The Minimum Spanning Tree Problem is a special case of the following optimization problem: Let E be a set of elements with a weight function w that assigns a real number to each element in E. For every $X \subseteq E$, let the weight of X be defined as

$$w(X) = \sum_{x \in X} w(x).$$

Let $\mathcal{I}$ be a set of subsets of E that satisfies properties (I1) and (I2) in Exercise 5.7(c). Find a maximal member of $\mathcal{I}$ of maximum weight.

The Greedy Algorithm (Algorithm 5.2.1) can be used to solve this problem. At each stage choose an element e of maximum weight, not previously chosen, so that $I \cup e \in \mathcal{I}$. Stop when no further such element can be found. Prove that $(E, \mathcal{I})$ is a matroid if and only if $\mathcal{I}$ satisfies:

(I1) $\phi \in \mathcal{I}$;

(I2) If $I_1 \in \mathcal{I}$ and $I_2 \subseteq I_1$, then $I_2 \in \mathcal{I}$; and

---

2 Lemma 5.2.2 implies that the set of spanning trees of a graph satisfies the two properties of bases. Thus a graph where E is the set of edges and $\mathcal{B}$ is the set of spanning trees is an example of a matroid (called a graphic matroid). Hassler Whitney defined matroids in his 1935 paper titled "On the abstract properties of linear dependence," and gave several examples including graphic matroids and vector matroids (Whitney, 1935). See Oxley (2011) for additional information.

(I3') For all weight functions w, the greedy algorithm produces a maximal member of $\mathcal{I}$ of maximum weight.[3]

5.9 The Bellman-Ford Algorithm, which is based on separate algorithms by Bellman (1958) and Ford and Fulkerson (1962), solves the single source shortest path problem for weighted digraphs, where some of the arc weights may be negative. Review this algorithm in (Cormen et al., 2001).

5.10 Finding the distance between every pair of vertices is known as the "All Pairs Shortest Paths problem." The Floyd-Warshall algorithm solves this problem for weighted graphs, where some of the edge weights may be negative. This is used to find the diameter of the graph. Floyd (1962) developed this algorithm based on a theorem by Warshall (1962). Review this algorithm in (Cormen et al., 2001).

5.11 Suppose we assign a label to each vertex in a connected graph in such a way that, given a starting vertex and the label assigned to the ending vertex, a shortest path to the ending vertex can be found only by looking at the labels in the neighborhood of each successive vertex in the path. That is, as we are constructing a shortest path, at each step we should be able to choose the next vertex on the path simply by comparing the labels on the neighbors of the current vertex to the label on the destination. In Graham and Pollak (1971) and Graham and Pollak (1972), the authors showed that this can be done for any connected graph. The labels they developed are made up of the symbols "0", "1" and "*". The choice of the next vertex in a shortest path depends on the Hamming distance between the label on a neighboring vertex and the label on the destination vertex. The Hamming distance between two lists of symbols of the same length is the number of positions in which the two lists differ. In this case, the symbols "0" and "1" are considered different, but "0" and "*" are considered the same, as are "1" and "*". Using their method, labels can be assigned to each vertex in such a way that if the neighbor with the smallest Hamming distance between its label and the destination's label is chosen, the resulting path will be a shortest path. Peter Winkler showed that for a graph with n vertices, labels of length $n - 1$ could be found which would work, (Winkler, 1983). Using the procedure described in his paper, find labelings for the graphs in Figures 3.3 and 3.4.

---

3 This amazing result by Borůvka in 1926 predates Whitney's 1935 paper introducing matroids and proves that the greedy strategy works for matroids and for nothing else. See Borůvka (1926a), Boruvka (1926b), and Nešetřil et al. (2001). Thus we say matroids are the objects that fully satisfy the greedy algorithm (Oxley (2011), p. 60).

6

Structure, Coloring, Higher Connectivity

This chapter takes us back to the origins of graph theory. Section 6.1 is on Eulerian circuits, cycle decompositions, and the optimization problem that arises from Eulerian circuits. Section 6.2 is on Hamiltonian cycles. Section 6.3 is on graph coloring and perfect graphs. Section 6.4 introduces higher connectivity and Section 6.5 is on Menger's theorem (Menger, 1927), which Diestel (2017) calls a "cornerstone" of graph theory.

6.1 Eulerian Circuits

Recall from Section 1.1 that a circuit that crosses every edge in the graph exactly once is called an Eulerian circuit. A graph with an Eulerian circuit is called an *Eulerian graph*. We may assume an Eulerian graph has $n \geq 3$ vertices and $m \geq 3$ edges since a graph with at most 2 vertices does not have a circuit. Thus the Königsberg Bridge problem from Chapter 1.1 can be rephrased as "Does the Königsberg graph have an Eulerian circuit?"

The next result appears in Euler (1736), the paper that began the subject, but the proof of sufficiency appears for the first time in Hierholzer (1873).

Theorem 6.1.1. (Euler's Theorem) *Let G be a connected graph with $m \geq 3$ edges. Then G has an Eulerian circuit if and only if every vertex has even degree.*

Proof. Suppose G has an Eulerian circuit C. Then C must leave a vertex as many times as it enters it. So every vertex has even degree.

Conversely, suppose every vertex in G has even degree. We must show that G has an Eulerian circuit. The proof is by induction on $m \geq 3$. The result holds if $m = 3$, since the only graph with even degree vertices and three edges is K_3, which is clearly Eulerian. Assume that the result holds for all graphs with even degree

Graphs and Networks, First Edition. S. R. Kingan.

vertices and fewer than $m \geq 4$ edges. Let C be the longest circuit in G. If C is an Eulerian circuit, then there is nothing to prove. Therefore suppose that there are edges in G not in C and consider $G \backslash C$, which is the subgraph obtained by deleting the edges of C from G and any isolated vertices.

First observe that every vertex in $G \backslash C$ has even degree. This is because every vertex in G has even degree and in removing edges of C we removed an even number of edges at each vertex. Second observe that $G \backslash C$ may be disconnected.

Let H be a connected component in $G \backslash C$. The number of edges in H is at least 3 and every vertex in H has even degree. By the induction hypothesis H has an Eulerian circuit D. Since G is connected, D has at least one vertex in common with C. So $C \cup D$ is a circuit in G that is longer than C. This is a contradiction to the maximality of C. Therefore C must be an Eulerian circuit. □

Note that Theorem 6.1.1 and its proof remain unchanged if the graph has parallel edges. An *Eulerian trail* is a trail that crosses every edge in the graph exactly once. The next result follows from Theorem 6.1.1.

Corollary 6.1.2. *Let G be a connected graph with $m \geq 1$ edges. Then G has an Eulerian trail from u to v if and only if every vertex of G except u and v has even degree.*

Proof. Suppose G has an Eulerian trail from u to v. For every vertex other than u and v, the trail has to leave a vertex as many times as it enters it. So all the vertices except u and v must have even degree.

Conversely, suppose all the vertices of G, except u and v, have even degree. Construct a new graph G' by adding vertex w and new edges wu and wv. Every vertex in G' has even degree and the number of edges is at least 3. Theorem 6.1.1 implies that G' has an Eulerian circuit C. Consequently $C - \{uw, vw\}$ is an Eulerian trail in G from u to v. □

In Hierholzer's proof circuits are "stitched" together to form the Eulerian circuit. The proof is constructive in nature and the technique gives a "stitching algorithm" for constructing the Eulerian circuit. Given a connected graph G with even degree vertices, Hierholzer's algorithm begins by finding a circuit C. If all the edges are used, then this is the Eulerian circuit. Otherwise consider the subgraph $G \backslash C$ obtained by deleting the edges of C and any isolated vertices. Find a circuit D in G / C and attach it to C. If all the edges are used, then $C \cup D$ is the Eulerian circuit, otherwise repeat. This process ends because the graph is connected. For a graph with $m \geq 3$ edges, Hierholzer's algorithm has complexity $O(m)$.

Figure 6.1 Hierholzer's algorithm.

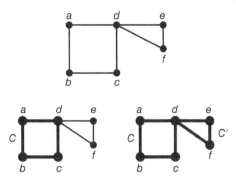

For example, consider the graph in Figure 6.1. When applying Hierholzer's algorithm the circuit $\{a, b, c, d\}$ may be obtained first, followed by circuit $\{d, e, f\}$. These two circuits are "stitched" together at vertex d to form the Eulerian circuit $\{a, b, c, d, e, f, d, a\}$.

Algorithm 6.1.3. Hierholzer's Algorithm

Input: A connected graph $G = (V, E)$ with even degree vertices.
Output: An Eulerian circuit C.

(1) Choose a vertex $v \in V$ and find an initial circuit C by traversing edges from v, avoiding edges already traversed, until returning to v.

(2) **While** $|C| < |E|$:
Choose a vertex u that is incident to C, but also incident to an edge not in C. (Since G is connected such a vertex must exist.)
Find a circuit D in $G \backslash C$ by traversing edges from u, avoiding edges already traversed, until returning to u. (Since G has even degree vertices, $G \backslash C$ will also have even degree vertices.)
Replace C with a new circuit formed by combining C and D at the common vertex u.

(3) **Return** C.

A *cycle decomposition* of a graph G is a set of edge-disjoint cycles whose union is $E(G)$. The next result is Oswald Veblen's characterization of Eulerian graphs in terms of a cycle decomposition (Veblen, 1912).[1] The proof is similar to the proof of Theorem 6.1.1.

---

1 Veblen wrote a topology book titled "Analysis Situs" and the first chapter was all about graphs (Veblen and Evans, 1922). This book appeared 16 years before König's book, which is considered to be the first graph theory textbook (König, 1936).

Proposition 6.1.4. *Let G be a connected graph with m ≥ 3 edges. Then G has a cycle decomposition if and only if every vertex has even degree.*

Proof. If G has a cycle decomposition, then the edge set is a disjoint union of cycles. So every vertex is in a cycle, and therefore every vertex has even degree.

Conversely, suppose every vertex in G has even degree. The proof is by induction on $m \geq 3$. Let C be a cycle in G. If $E(G) = C$, then there is nothing to prove. Therefore suppose that there are edges in G not in C. Let $G \backslash C$ be the graph obtained by deleting the edges of C from G and any isolated vertices. As in the proof of Theorem 6.1.1, every vertex of $G \backslash C$ has even degree and $G \backslash C$ may be disconnected. Let $H_1, \ldots H_k$ be the connected components of $G \backslash C$. Every vertex in H_i, where $1 \leq i \leq k$ has even degree, and therefore H_i has at least 3 edges. By the induction hypothesis $E(H_i)$ is the disjoint union of cycles. Since G is connected, each component has at least one vertex in common with another component. Therefore $G \backslash C$ has a set of edge-disjoint cycles whose union is $E(G \backslash C)$. This set of edge-disjoint cycles together with C form a set of edge-disjoint cycles whose union is $E(G)$. □

A *cycle double cover* of a graph is a set of cycles C such that every edge of the graph appears in two cycles of C. Note that the cycles may be repeated. For example, if $G = C_n$, then the same cycle is used twice to conclude that every edge is in two cycles. A graph with a cycle decomposition also has a cycle double cover for the same reason. An obvious necessary condition for the existence of a cycle double cover is that the graph has no bridges. Szekeres (1973) and Seymour (1980) conjectured that this is also a sufficient condition.

Conjecture 6.1.5. (Cycle Double Cover Conjecture) Let G be a connected bridgeless graph. Then G has a cycle double cover.

Subsequently, Bondy (1990) made a related conjecture.

Conjecture 6.1.6. (Small Cycle Double Cover Conjecture) Let G be a connected bridgeless graph with $n \geq 3$ vertices. Then G has a cycle double cover with at most $n - 1$ cycles.

Although the notion of maxima and minima permeated Euler's work, he did not think of asking an optimization question related to Eulerian circuits. This had to wait until 1960 when Chinese mathematician Mei-Ko Kwan asked what is the minimum length of a closed walk that crosses every edge at least once in a graph that is not necessarily Eulerian (Kwan, 1962). Specifically, he presented the problem as follows [2]:

> A postman has to deliver letters to a given neighborhood. He needs to walk through all the streets in the neighborhood and back to the post-office. How can he design his route so that he walks the shortest distance?

This is called the Postman Problem or more generally the Route Inspection Problem since the same situation arises in many different contexts such as street-sweeping, RNA chains, and de Bruijn sequences (Roberts, 2005).

The streets in a town may be viewed as a graph with the intersections of the streets forming vertices and the streets forming edges. In the Postman Problem three simplifying assumptions are made:

(1) The graph is undirected;
(2) The post office is at a vertex; and
(3) There are no streets with houses on both sides.

If the graph is Eulerian, then the postman will follow an Eulerian circuit beginning and ending at the post office. If the graph is not Eulerian, then she will have to walk along some streets at least twice. The question becomes what is the fewest number of edges that must be doubled to make the graph Eulerian. This is called *Eulerizing* a graph. Euler came close to posing this problem when he noted that a graph that is not Eulerian can be made Eulerian by doubling all the edges because when doing so every vertex has even degree.

Let G be a graph with $m \geq 3$ edges. Clearly m is a lower bound (which occurs when the graph is Eulerian) and $2m$ is an upper bound (which occurs when all the edges are doubled). The upper bound is sharp. The number of edges required to Eulerize a tree with m edges is $2m$ since every bridge must be doubled, and in a tree every edge is a bridge. The first graph in Figure 6.2 has four odd degree vertices. It can be Eulerized by doubling four edges. The second graph has six odd degree vertices, two pairs of which are adjacent. It can also be Eulerized by doubling four edges. In both cases these are the minimum number of edges that must be added to Eulerize the graph.

Kwan gave a sufficient condition for Eulerizing a graph (Kwan, 1962), which we will state without a proof. He also posed a similar problem for strongly connected digraphs and for weighted digraphs. Edmonds and Johnson gave an efficient algorithm by showing that the Postman Problem is equivalent to finding a maximum weight matching; a topic in Chapter 8 (Edmonds and Johnson, 1973).

---

2 See the survey article "Euler, Mei-Ko Kwan, Königsberg, and a Chinese postman" by Grötschel and Yuan (2012) and the book *Covering Walks in Graphs* by Fujie and Zhang (2014).

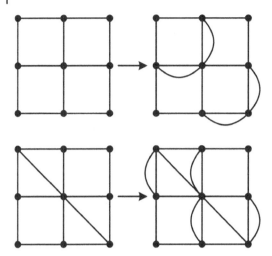

Figure 6.2   Eulerizing graphs.

Theorem 6.1.7. (Kwan's Theorem)   *Suppose G is a connected graph. If at most half the edges in each cycle are doubled, then G is Eulerian.*

Finally, a natural optimization question to ask about an Eulerian graph G with $n \geq 3$ vertices is "What is the minimum number of cycles in a cycle decomposition of G?" If G has a vertex v such that $deg(v) = n - 1$, then since any cycle through v has exactly two edges incident to v, a cycle decomposition must have at least $\lfloor \frac{n-1}{2} \rfloor$ cycles. György Hajós conjectured that this is also an upper bound (Lovász, 1968).

Conjecture 6.1.8. (Hajós' Cycle Conjecture)   Let G be an Eulerian graph with $n \geq 3$ vertices. The number of cycles in a cycle decomposition is at most $\lfloor \frac{n-1}{2} \rfloor$.

Suppose G is an Eulerian graph and Conjecture 6.1.8 holds for G. Then G has a cycle decomposition with at most $\lfloor \frac{n-1}{2} \rfloor$ cycles. By taking two copies of such a cycle decomposition, we obtain a cycle double cover of G with at most $(n - 1)$ cycles. Thus for Eulerian graphs proving Conjecture 6.1.8 would also prove Conjecture 6.1.6.

6.2   Hamiltonian Cycles

Recall from Section 1.1 that a Hamiltonian cycle is a cycle that contains every vertex of a graph exactly once. A graph with a Hamiltonian cycle is called a Hamiltonian graph. Figure 6.3 displays the dodecahedron graph and $K_{3,3}$, both of which have Hamiltonian cycles (shown in bold). To prove that a graph has a Hamiltonian cycle, we must display the cycle. This is no small task as the graph gets large. There

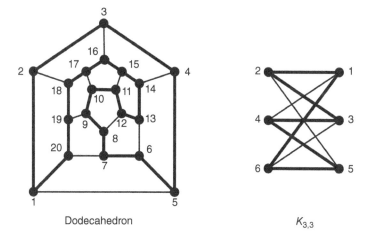

Figure 6.3 Hamiltonian graphs.

are several sufficient conditions for the existence of a Hamiltonian cycle and several necessary conditions, but a necessary and sufficient condition remains elusive. Finding a Hamiltonian cycle is one of the NP-complete problems in (Karp, 1972).

The study of Hamiltonian cycles was initiated by Thomas Kirkland in the mid-nineteenth century. However, it was William Hamilton who made it famous and therefore, perhaps erroneously, these cycles are called Hamiltonian cycles. Kirkland was a rector of a small parish in England, and among the mathematical objects he studied were polyhedra. A *polyhedron* is a 3-dimensional object bounded by a finite number of polygons (faces). The polygons meet in straight lines (edges) and the lines meet in points (vertices).

In an 1856 paper Kirkland asked if one could find a cycle that passes through each vertex of a polyhedron exactly once (Kirkman, 1856). Shortly thereafter, Hamilton asked the same question of the dodecahedron graph in the form of a game that he called the Icosian Game (Hamilton, 1858). He patented this game and marketed it widely. Another version of his game was called "The Traveler's Dodecahedron." The 20 vertices of the dodecahedron represented 20 cities: Brussels, Canton, Delhi, and so on, ending with Zanzibar. Each vertex had a peg. The challenge was to pass a thread around each peg exactly once to obtain a cycle covering all the pegs, thereby taking a voyage around the world (Biggs et al., 1976).

We will begin with two necessary conditions for having Hamiltonian cycles. Clearly, the complete graph K_n has a Hamiltonian cycle and $K_{3,3}$ has a Hamiltonian cycle (shown in Figure 6.3). The next result is a straightforward necessary condition for the existence of Hamiltonian cycles in bipartite graphs.

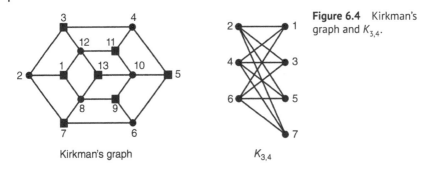

Figure 6.4 Kirkman's graph and $K_{3,4}$.

Kirkman's graph

$K_{3,4}$

Proposition 6.2.1. *Let G be a connected bipartite graph. If G has a Hamiltonian cycle, then G has an even number of vertices.*

Proof. Let $V(G) = X \cup Y$ and C be the Hamiltonian cycle in G. Since G is bipartite, this cycle must visit X and Y alternatively and return to the starting vertex. Thus there must be an even number of vertices in G. □

The first graph in Figure 6.4 appears in Kirkman's paper. It is not obviously bipartite, but a careful look indicates that the two vertex classes are the vertices drawn with circles and the vertices drawn with squares (Biggs et al., 1976). Since it has an odd number of vertices, it is non-Hamiltonian by Proposition 6.2.1. For the same reason the second graph in Figure 6.4 is also non-Hamiltonian. A *Hamiltonian path* is a path that contains every vertex. Both non-Hamiltonian graphs in Figure 6.4 have a Hamiltonian path (follow the numbered vertices in order).

The converse of Proposition 6.2.1 is not true. For example, any bipartite graph with a leaf (vertex of degree 1) is non-Hamiltonian even if the number of vertices is even.

The next proposition gives another necessary condition for finding a Hamiltonian cycle.

Proposition 6.2.2. *Let G be a connected graph. If G has a Hamiltonian cycle, then for every subset S of vertices, the number of connected components in $G - S$ is at most $|S|$.*

Proof. Let C be a Hamiltonian cycle of G and let $G_1, \ldots, G_k$ be the components of $G - S$. For each $i \in \{1, \ldots, k\}$, when C leaves G_i the next vertex of C belongs to S. Therefore $k \leq |S|$. □

Hamiltonian cycles and Eulerian circuits are not necessarily related. See the two non-Hamiltonian graphs in Figure 6.5. The first is Eulerian and the second is non-Eulerian. The cycle graph C_n is both Hamiltonian and Eulerian. So is

Figure 6.5 Non-Hamiltonian graphs.

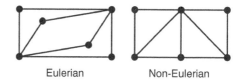

Eulerian   Non-Eulerian

the complete graph K_n with n odd. However, K_n with n even is Hamiltonian, but non-Eulerian. The complete bipartite graph $K_{r,s}$ with r and s both even is Hamiltonian and Eulerian, whereas $K_{r,s}$ with r and s both odd is Hamiltonian, but non-Eulerian.

Next, we will look at sufficient conditions for Hamiltonian cycles. Dirac (1952) proved a sufficient condition for Hamiltonian cycles based on vertex degrees. Intuitively, if vertices have high degrees, then there will be more edges in the graph and the likelihood of finding a Hamiltonian cycle is higher. The question is how high? Dirac proved that if the degree of every vertex is at least half the number of vertices, then G has a Hamiltonian cycle. Subsequently, Ore (1960) strengthened Dirac's result. The next result is Ore's theorem, and Dirac's theorem is obtained as a corollary. The proof is taken from Chartrand et al. (2011).

A non-Hamiltonian graph G is called *maximal non-Hamiltonian* if $G + uv$ is Hamiltonian for every pair of non-adjacent vertices u and v. In other words adding one more edge results in a Hamiltonian graph.

Theorem 6.2.3. (Ore's Theorem)   *Let G be a connected graph with $n \geq 3$ vertices. If*

$$deg(u) + deg(v) \geq n$$

for every pair of non-adjacent vertices u and v, then G has a Hamiltonian cycle.

Proof. Suppose the theorem is not true. Let G be a maximal non-Hamiltonian graph that satisfies the hypothesis of the theorem. Choose non-adjacent vertices u and v and join them by an edge to obtain $G + uv$. By construction $G + uv$ has a Hamiltonian cycle and this cycle must use edge uv. We may write this cycle as

$$uu_1u_2 \cdots u_{n-2}vu.$$

First, observe that for $2 \leq i \leq n - 2$, if uu_i is an edge in $G + uv$, then vu_{i-1} is not an edge in $G + uv$. To confirm this suppose it is not true and uu_i and vu_{i-1} are both edges in $G + uv$, for some i. Then

$$uu_1u_2 \cdots u_{i-1}vu_{n-2} \ldots u_iu$$

is a Hamiltonian cycle in G as shown in the following diagram. This is a contradiction since we began by assuming that G has no Hamiltonian cycle.

Let t be the number of vertices among $u_2, \ldots, u_{n-2}$ joined by an edge to u in $G +$ uv. Since uv and uu_1 are edges, $deg_{G+uv}(u) = t + 2$. By the previous observation, if uu_i is an edge, then vu_{i-1} is not an edge, and since uv and vu_{n-2} are edges,

$$deg_{G+uv}(v) \le (n-3) - t + 2.$$

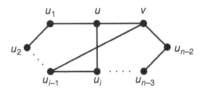

Therefore

$$deg_{G+uv}(u) + deg_{G+uv}(v) \le (t+2) + (n-3-t+2) = n+1.$$

When edge uv is removed it reduces the degrees of u and v by 1. Consequently

$$deg_G(u) + deg_G(v) \le n - 1.$$

This is a contradiction to the hypothesis. Therefore G is Hamiltonian. □

Corollary 6.2.4. (Dirac's Theorem) *Let G be a connected graph with $n \ge 3$ vertices. If $deg(v) \ge \frac{n}{2}$ for every vertex v, then G has a Hamiltonian cycle.*

Proof. If every vertex has degree at least $\frac{n}{2}$, then for every pair of vertices u and v,

$$deg(u) + deg(v) \ge \frac{n}{2} + \frac{n}{2} = n.$$

So G has a Hamiltonian cycle by Theorem 6.2.3. □

Bondy and Chvátal (1976) developed a slightly different approach to obtaining a sufficient condition for Hamiltonian cycles. They began by noting that Ore had in fact proved a slightly stronger statement than the statement of Theorem 6.2.3.

Proposition 6.2.5. *Let G be a connected graph with $n \ge 3$ vertices. Let u and v be non-adjacent vertices such that $deg(u) + deg(v) \ge n$. Then $G + uv$ has a Hamiltonian cycle if and only if G has a Hamiltonian cycle.*

Proof. Suppose $G + uv$ has a Hamiltonian cycle C. If uv is not an edge in C, then C is a Hamiltonian cycle in G. Therefore suppose that uv is an edge in cycle C and $C = uu_1u_2 \cdots u_{n-2}vu$. By the same argument as in the proof of Theorem 6.2.3, for $2 \leq i \leq n - 2$, if uu_i is an edge in $G + uv$, then vu_{i-1} is not an edge in $G + uv$. Consequently $deg_G(u) + deg_G(v) \leq n - 1$, which is a contradiction.

The converse is immediate. If G has a Hamiltonian cycle, then the graph obtained by linking a pair of non-adjacent vertices also has a Hamiltonian cycle.

□

Let G be a connected graph with n vertices. The *closure of G*, denoted by *closure(G)*, is the graph obtained by recursively joining pairs of non-adjacent vertices whose degrees add up to n or greater until no such vertices remain. In other words construct a new graph G_1 from G by adding edges uv for each pair of vertices u and v with $deg_G(u) + deg_G(v) \geq n$.

Repeat this process for G_1 to obtain G_2, and continue as long as possible. Clearly it ends in a finite number of steps, either at K_n or at a graph with fewer edges than K_n. For example, in Figure 6.6, the closure of G is not a complete graph, but the closure of H is the complete graph K_6. Since the complete graph has a Hamiltonian cycle, we may conclude that if *closure(G)* is a complete graph, then G is Hamiltonian, thereby obtaining another necessary condition for Hamiltonian cycles.

Corollary 6.2.6. *Let G be a connected graph with $n \geq 3$ vertices. If the closure of G is complete, then G is Hamiltonian.*

There is a large body of work on necessary conditions and sufficient conditions for Hamiltonian cycles. See the survey articles Gould (1991) and Gould (2003).

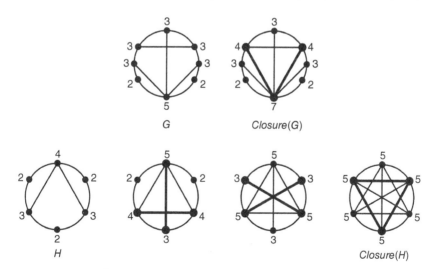

Figure 6.6 Closure of a graph.

6.3 Coloring

A *proper coloring* of a graph G is an assignment of colors to the vertices so that no two adjacent vertices get the same color. A proper coloring in which t colors are used is called a *t-coloring* and the graph is called *t-colorable*. The *chromatic number* of G, denoted by $\chi(G)$, is the smallest number t such that G has a t-coloring.

For example, the graph in Figure 6.7 can be colored with 3, 4, or 5 colors, but 3 is the minimum. So $\chi(G) = 3$. The chromatic number of the complete graph K_n is $n - 1$, since each vertex is joined to the other $n - 1$ vertices. The chromatic number of a bipartite graph is 2, and since trees and even cycles are bipartite, their chromatic numbers are also 2. However, the chromatic number of an odd cycle is 3.

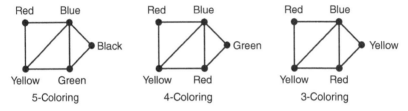

Figure 6.7 Graph coloring.

Instead of assigning vertices actual colors, we may simply assign them numbers as shown in the following diagram. It helps to think of a coloring as a partition of the vertex set into disjoint subsets so that no two adjacent vertices are in the same subset. Color is proxy for anything that partitions the vertex set in this manner. For this example in Figure 6.8, $\{\{v_1, v_3\}, \{v_5\}, \{v_2, v_4\}\}$ is a partition of the vertex set. In general, if $\chi(G) = t$, then G is a t-partite graph.

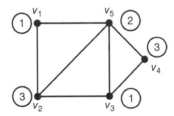

An obvious upper bound for the chromatic number is the number of vertices, but we can do better. Recall that $\Delta(G)$ is the largest degree of the vertices.

Proposition 6.3.1. *Let G be a graph with $n \geq 1$ vertices. Then $\chi(G) \leq \Delta(G) + 1$.*

Proof. The proof is by induction on the number of vertices. The result holds for the trivial graph. Assume that the result is true for graphs with $n - 1$ vertices and let G be a graph with $n \geq 2$ vertices. Let v be any vertex in G and consider $G - v$. By the induction hypothesis

$$\chi(G - v) \leq \Delta(G - v) + 1 \leq \Delta(G) + 1.$$

Since v has at most $\Delta(G)$ neighbors, there must be at least one unused color for v. Therefore $\chi(G) \leq \Delta(G) + 1$. □

The bound in Proposition 6.3.1 is tight. Equality holds for C_n, where n is odd since $\chi(C_n) = 3$ and $\Delta(C_n) = 2$. Equality also holds for K_n since $\chi(K_n) = n$ and $\Delta(K_n) = n - 1$. As it turns out these are the only two exceptions according to the next theorem by Brooks (1941). The proof is taken from Melnikov and Vizing (1969) and uses what is known as a "Kempe chain argument."

Let G be a graph with a t-coloring $1, 2, \ldots, t$. A *Kempe chain* in G with respect to the t-coloring is a subgraph induced by all the vertices that receive one of two specified colors i and j. This induced subgraph H_{ij} may be disconnected, but if H_{ij} is connected, then we can exchange colors i and j to get another t-coloring of G.

Theorem 6.3.2. (Brook's Theorem)   *Let G be a connected graph that is neither an odd cycle nor a complete graph. Then $\chi(G) \leq \Delta(G)$.*

Proof. Suppose G is a connected graph and G is not an odd cycle or a complete graph. If $\Delta(G) = 2$, then G is a path or an even cycle and the result holds. We may assume for the rest of the proof that $\Delta(G) \geq 3$ and the number of vertices $n \geq 4$.

The proof is by induction on n. The result holds for graphs on 4 vertices. Assume the result is true for all graphs with fewer than n vertices and suppose G has n vertices. First suppose that G has a vertex v, where $deg(v) < \Delta(G)$. By the induction hypothesis $G - v$ is an odd cycle, a complete graph, or

$$\chi(G - v) \leq \Delta(G - v) = \Delta(G).$$

If $G - v$ is an odd cycle, then G is an odd cycle with an additional vertex v joined to at most 2 other vertices. In this case $\chi(G) = \Delta(G) = 3$. If $G - v \cong K_{n-1}$, then since $G \ncong K_n$, G must be K_{n-1} with an additional vertex v joined to at most $n - 2$ other vertices. In this case $\chi(G) = \Delta(G) = n$. In the third situation $G - v$ has a coloring with at most $\Delta(G)$ colors. However, since $deg(v) < \Delta(G)$, fewer than $\Delta(G)$ colors are used by the neighbors of v. One of the unused colors may be applied to v to obtain a coloring of G with at most $\Delta(G)$ colors.

Therefore we may assume that G is a regular graph of degree t, where $3 \leq t \leq n - 2$. Let v be a vertex in G and consider $G - v$. By the induction hypothesis $G - v$ is an odd cycle, a complete graph, or

$$\chi(G - v) \leq \Delta(G - v) = t.$$

If $G - v$ is an odd cycle, then G is not regular; a contradiction. If $G - v$ is a complete graph, then G is also a complete graph; a contradiction. So we may assume that $\chi(G - v) \leq t$.

Let $N(v) = \{v_1, \ldots, v_t\}$. If the t neighbors of v use less than t colors, then one of the unused colors may be applied to v to obtain a t-coloring of G, and we are done.

Therefore suppose $v_1, \ldots, v_t$ are colored with the t available colors $1, \ldots, t$. If v_i has two or more neighbors colored with the same color, then v_i may be colored the missing color among its neighbors. So color i becomes available for v, and we get a t-coloring of G.

Select two neighbors v_i and v_j colored i and j, respectively, and consider the Kempe chain H_{ij} with respect to the t-coloring. If v_i and v_j are in different components of H_{ij}, then there is no path from v_i to v_j in H_{ij}. Recolor v_i with color j and exchange colors i and j in the component containing v_i. This frees color i, which may be assigned to v to obtain a t-coloring of G.

Therefore we may assume that v_i and v_j are in the same component of H_{ij}. Then there is a path from v_i to v_j with vertices colored alternately i and j. Suppose there is a vertex $x \neq v_i$ colored i along this path with three or more neighbors in H_{ij} all colored j. Then there is at least one unused color besides i and j among neighbors of x in G, which may be assigned to x. Once this assignment is made, v_i and v_j will be in different components of H_{ij} and the argument from the previous paragraph may be applied.

Thus we may assume that H_{ij} is precisely a path from v_i to v_j. Consider vertex v_i colored i and another vertex v_k colored k, where $k \neq j$, and the Kempe chain H_{ik}. As shown in the previous paragraph, H_{ik} is also a path from v_i to v_k. Suppose, if possible, H_{ij} and H_{ik} intersect in a vertex $y \neq v_i$. Then y is colored i. Moreover, y has at least four neighbors that use only two colors j and k. So there is at least one unused color among neighbors of y in G distinct from j and k. This color may be assigned to y, and this puts v_i and v_j in different components of H_{ij} and v_i and v_k in different components of H_{ik}. Therefore we may assume that two distinct Kempe chains H_{ij} and H_{ik} have only vertex v_i in common.

Let z be a vertex distinct from v_i and v_j along a path between v_i and v_j in G. Then z is also on a path between v_i and v_j in $G - v$. If z has received color k distinct from i and j in the t-coloring of $G - v$, then z is in the Kempe chains H_{ij} and H_{ik}, which is impossible. So in $G - v$ either z receives color i or j and lies in the Kempe chain H_{ij} or v_i and v_j are adjacent in $G - v$. Thus $G - v$ is a path of even length or $G - v$ is a complete graph. In the former situation G is an odd cycle and in the latter situation G is a complete graph; a contradiction. □

Finding the chromatic number of a graph is an NP-complete problem (Karp, 1972). Practically speaking, however, a greedy-type algorithm can quickly find a coloring, although not necessarily a minimum coloring.

Algorithm 6.3.3. Greedy Coloring Algorithm

Input: A connected graph G with vertices labeled $v_1, \ldots, v_n$.
Output: A coloring of G with at most $\Delta(G) + 1$ colors.

(1) Color v_1 with color 1.
(2) For vertices labeled $v_2, \ldots, v_n$ color v_i with the lowest numbered color that has not been used by any of its neighbors. If all the colors are used, then assign v_i a new color.

Proposition 6.3.4. *Algorithm 6.3.3 finds a coloring of G with at most $\Delta(G) + 1$ colors.*

Proof. Let $v_1, \ldots, v_n$ be a labeling of the vertices of G and let the colors be represented by numbers $1, \ldots, n$. For $1 \leq i \leq n$, assign v_i the lowest number not used for its neighbors among $v_1, \ldots, v_{i-1}$. Since v_i has at most $\Delta(G)$ neighbors, a color among the $\Delta(G) + 1$ colors is always available. Thus G may be colored by at most $\Delta(G) + 1$ colors. □

The computational complexity of Algorithm 6.3.3 for a graph with n vertices and m edges is $O(n + m)$. Observe that Proposition 6.3.4 provides an alternate proof of Proposition 6.3.1; one that has an algorithmic flavor.

Algorithm 6.3.3 does not necessarily give a coloring with fewest colors. The coloring obtained depends heavily on the ordering of vertices. Figure 6.8 shows the path P_4 with two different vertex labels. For the first labeling Algorithm 6.3.3 assigns v_1 color 1, v_2 color 2, v_3 color 1, and v_4 color 2, thereby obtaining a 2-coloring. However, for the second labeling, Algorithm 6.3.3 assigns v_1 color 1, v_2 color 1, v_3 color 2 and v_4 color 3, thereby obtaining a 3-coloring. Figure 6.8 also shows the graph obtained from $K_{4,4}$ with 4 edges removed (shown dotted) with two different vertex labels. For the first labeling, Algorithm 6.3.3 assigns v_1, v_2, v_3, v_4 color 1 and v_5, v_6, v_7, v_8 color 2, giving a 2-coloring. For the second labeling, Algorithm 6.3.3 assigns v_1 and v_2 color 1, v_3 and v_4 color 2, v_5 and v_6 color 3, and v_7 and v_8 color 4, giving a 4-coloring.

We can define edge coloring in a manner similar to vertex coloring. A *proper edge coloring* of a graph G is an assignment of colors to the edges so that no two adjacent edges have the same color. A proper edge coloring in which t colors are used is called a *t-edge coloring*. The *edge chromatic number* of G, denoted by $\chi'(G)$, is the smallest number t such that G has a t-edge coloring.

For example, $\chi'(K_n) = n - 1$ since each of the $n - 1$ edges incident to a vertex requires a different color. Clearly $\chi'(P_n) = 2$. If C_n is an even cycle, then $\chi'(C_n) = 2$ since every alternate edge can receive the same color. If C_n is an odd cycle, then $\chi'(C_n) = 3$. An obvious lower bound for edge chromatic number is $\Delta(G)$. The next result, stated without proof, gives a tight upper bound (Vizing, 1964).

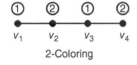

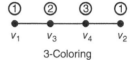

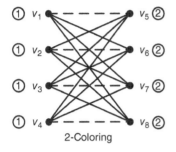

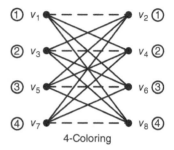

Figure 6.8 Greedy Coloring Algorithm.

Theorem 6.3.5. (Vizing's Theorem) *Let G be a connected graph. Then $\Delta(G) \leq \chi'(G) \leq \Delta(G) + 1$.*

The *clique number* of a graph G, denoted by $\omega(G)$, is the size of a maximum clique in G. Since every vertex in a clique must have a different color

$$\chi(G) \geq \omega(G) \geq 2.$$

A graph G is called *perfect* if $\chi(H) = \omega(H)$ for every induced subgraph H of G. In the definition of perfect graphs we require equality to hold, not only for G, but also for every induced subgraph of G in order to avoid trivial examples. A disconnected graph with two components, one of which is a large complete graph and the other with a few edges, satisfies $\chi(G) = \omega(G)$, but this is not an interesting example.

Examples of perfect graphs include bipartite graphs since the chromatic number of a bipartite graph is 2, and every induced subgraph of a bipartite graph is also bipartite. Similarly trees are perfect since trees are bipartite. An even cycle is perfect, however, an odd cycle of size greater than 3 is not perfect, since it has chromatic number 3 and clique number 2. The class of perfect graphs is closed under induced subgraphs, so any graph that has an induced odd cycle of size greater than 3 is not perfect.

Claude Berge[3] introduced perfect graphs and among the many conjectures that he made, two stand out (Berge, 1963). As noted in the previous paragraph, $K_{r,s}$

---

3 Berge wrote a popular book on graph theory (Berge, 1958) and in the preface of the English translation Gian-Carlo Rota wrote "Two Frenchmen have played a major role in the renaissance of combinatorics: Berge and Schutzenberger. Berge has been the more prolific writer, and his

where $r \leq s$, is a perfect graph. Its complement $\overline{K}_{r,s}$ consists of two disconnected components, a clique of size r and a clique of size s. So $\chi(\overline{K}_{r,s}) = s = \omega(\overline{K}_{r,s})$. Consequently $\overline{K}_{r,s}$ is also perfect. Berge conjectured that the complement of any perfect graph is perfect. This conjecture, known as the Perfect Graph Conjecture, was settled by Lovász (1972). It is stated here without a proof.

Theorem 6.3.6. (Perfect Graph Theorem) *A graph is perfect if and only if its complement is perfect.*

Second, Berge conjectured that a graph G is perfect if and only if neither G nor $\overline{G}$ contain an odd cycle of size greater than 3. This conjecture, called the "Strong Perfect Graph Conjecture," was settled by Chudnovsky et al. (2006). An induced odd cycle of size at least 5 is called an *odd hole* and an induced subgraph that is the complement of an odd hole is called an *antihole*.

Theorem 6.3.7. (Strong Perfect Graph Theorem) *A graph is perfect if and only if it has no odd hole or antihole.*

6.4 Higher Connectivity

The connectivity described in Section 1.1 is just the first level of connectivity. The definition of a connected graph essentially states that at least one vertex must be removed to disconnect the graph. In some graphs several vertices have to be removed to disconnect the graph. It makes sense to define levels of connectivity in terms of the smallest number of vertices that must be removed to disconnect the graph. There is a small technicality to consider. The most linked a graph can be is K_n, so we want the connectivity of K_n to be $n - 1$. However, if we remove $n - 1$ vertices from K_n, then we get the graph with one vertex which is trivially connected. We must take this into account in our definition of higher connectivity.

Let G be a connected graph. A set of vertices whose removal disconnects G or results in the trivial graph is called a *vertex-cut*. The *connectivity* of G, denoted by $\kappa(G)$, is the size of the smallest vertex-cut. For $k \geq 1$, we say G is *k-connected* if $\kappa(G) \geq k$. Consider the following examples:

(1) For $n \geq 2$, $\kappa(P_n) = 1$ since removing one vertex disconnects a path. So P_n is 1-connected (or connected).

books have carried the word farther and more effectively that anyone anywhere. I recall the pleasure of reading the disparate examples in his first book, which made it impossible to forget the material. Soon after reading, I would be one of many who unknotted themselves from the tentacles of the continuum and joined the Rebel Army of the Discrete." See also Berge's recollections "Motivations and history of some of my conjectures" (Berge, 1997).

(2) For $n \geq 3$, $\kappa(C_n) = 2$ since at least two vertices must be removed to disconnect a cycle. A cycle is 2-connected.

(3) For $n \geq 4$, $\kappa(W_{n-1}) = 3$ since at least 3 vertices must be removed to disconnect a wheel. A wheel is 3-connected.

A set of edges whose removal disconnects G is called an *edge-cut*. The *edge-connectivity* of a graph G, denoted by $\lambda(G)$, is the size of the smallest edge-cut. For $k \geq 1$, we say G is *k-edge-connected* if $\lambda(G) \geq k$. For example, $\lambda(P_n) = 1$, $\lambda(C_n) = 2$, $\lambda(W_{n-1}) = 3$, and $\lambda(K_n) = n - 1$.

The next proposition gives a relationship between vertex and edge connectivity.

Proposition 6.4.1. *Let G be a connected graph. Then* $\kappa(G) \leq \lambda(G) \leq \delta(G)$.

Proof. Clearly $\lambda(G) \leq \delta(G)$ since removing edges incident to a vertex of minimum degree will disconnect the graph. Suppose S is an edge-cut of size $\lambda(G)$ and T is the set of vertices formed by selecting one vertex incident to each edge in S. Then removing the vertices in T will disconnect G. So T is a vertex-cut. Since $\kappa(G)$ is the size of the smallest vertex-cut

$$\kappa(G) \leq |T| = |S| = \lambda(G).$$

$\square$

For a fixed value of $\kappa(G)$, we can make the difference between $\kappa(G)$ and $\lambda(G)$ as large as desired. In Figure 6.9, $\kappa(G_1) = 1$ and $\lambda(G_1) = 3$. Observe that G_1 is the clique-sum of two copies of K_4 attached at a vertex. Suppose G is the clique-sum of two copies of K_n attached at a vertex, then $\kappa(G) = 1$ and $\lambda(G) = n - 1$.

The second graph in Figure 6.9 is the clique sum of two copies of K_5 attached at an edge. In this case $\kappa(G_2) = 2$ and $\lambda(G_2) = 4$. If G is the clique-sum of two copies of K_n attached at an edge, then $\kappa(G) = 2$ and $\lambda(G) = n - 1$. Continuing this way by letting G be the clique sum of two copies of K_n attached across a clique K_t, where $t < n$, we can obtain a graph with any combination of $\kappa(G)$ and $\lambda(G)$.

We can also make $\lambda(G)$ and $\delta(G)$ differ as much as we want. In Figure 6.9 $\kappa(G_3) = 2$, $\lambda(G_3) = 4$, and $\delta(G_3) = 5$. Replacing K_6 with a larger K_n maintains $\lambda(G)$ at 4, but increases $\delta(G)$.

The next corollary gives a lower bound on the number of edges in a graph with a specified edge connectivity.

Corollary 6.4.2. *Let G be a graph with $n \geq 3$ vertices and m edges. Then* $m \geq \lceil \frac{n\lambda(G)}{2} \rceil$.

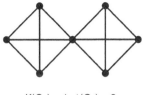

$$K(G_1) = 1, \lambda(G_1) = 3$$

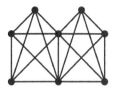

$$K(G_2) = 2, \lambda(G_2) = 4$$

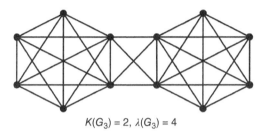

$$K(G_3) = 2, \lambda(G_3) = 4$$

$$K(G_4) = 2, \lambda(G_4) = 2$$

Figure 6.9 Vertex and edge connectivity.

Proof. Proposition 6.4.1 implies that $\lambda(G) \leq \delta(G) \leq deg(v)$ for every vertex v. Therefore

$$2m = \sum deg(v) \geq n\lambda(G)$$

and

$$m \geq \left\lceil \frac{n\lambda(G)}{2} \right\rceil.$$

$\square$

The converse of Corollary 6.4.2 is not true. That is, $m \geq \lceil \frac{nt}{2} \rceil$ does not imply $\lambda(G) = t$. For example, consider the fourth graph in Figure 6.9. In this graph $n = 6$ and $m = 9$, so $m \geq \frac{6 \times 3}{2}$, but $\lambda(G) \neq 3$. Moveover, Corollary 6.4.2 implies that

$$m \geq \left\lceil \frac{n\kappa(G)}{2} \right\rceil,$$

since by Proposition 6.4.1, $\lambda(G) \geq \kappa(G)$.

A graph that is 1-connected, but not 2-connected, has a cut vertex. Recall from Section 2.2 a block in a 1-connected graph is a maximal induced subgraph with no cut vertices. So a block is either a 2-connected graph or a single edge that is a bridge. Thus $\kappa(G) = 1$ if and only if G is the clique sum of its blocks attached at cut vertices. The next set of results gives equivalent conditions for 2-connected graphs and 2-edge-connected graphs.

Proposition 6.4.3. *A graph is 2-edge-connected if and only if every edge lies on a cycle.*

Proof. Suppose G is 2-edge-connected. Then for every edge $e = uv$, $G\backslash e$ is connected. This means that although e is deleted, there is a $u - v$ path P in $G\backslash e$, and therefore $P \cup e$ forms a cycle in G. Thus every edge of G lies on a cycle. The converse argument reverses the direction of the implications. □

Theorem 6.4.4. *A graph is 2-connected if and only if every pair of vertices lies on a common cycle.*

Proof. Suppose G is 2-connected. Then Proposition 6.4.1 implies that G is 2-edge-connected and Proposition 6.4.3 implies that every edge lies on a cycle. Let $v \in V(G)$ and suppose, if possible, there is a vertex w that does not lie on a cycle with v. Then $w \notin N(v)$ since all the vertices adjacent to v lie on a cycle with v. Since G is connected, there is a $v - w$ path P in G. Denote the vertices of P in order as

$$v = x_1, x_2, \ldots, x_k = w,$$

where $k \geq 3$ (since w is not adjacent to v). Let i be the smallest integer such that v and x_{i-1} lie on a cycle C, but v and x_i do not lie on a cycle. Since G is 2-connected, x_{i-1} is not a cut vertex. So there is a path P' between v and x_i that does not contain x_{i-1}. There are two cases as shown in the following diagram:

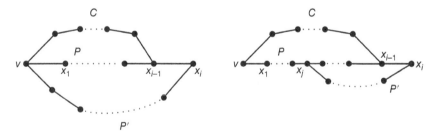

In the first case, path P' meets path P only in v (see the first figure in the diagram). Then going from v to x_i along P and from x_i to v along P' gives a cycle containing both v and x_i.

In the second case, path P' meets path P at some vertex x_j, where $j \leq i - 2$ (see the second figure in the previous diagram). A cycle containing v and x_i can be constructed by going from v to x_{i-1} along C, followed by edge $x_{i-1}x_i$, then going from x_i to x_j along P', and finally going from x_j to v along P. Thus in both cases v and x_i lie on a cycle; a contradiction to our assumption.

Conversely, suppose every pair of vertices lie on a cycle. Suppose, if possible, G is not 2-connected. Then G has a cut vertex v, and $G - v$ is disconnected. Select a pair of vertices x, y in different components of $G - v$. Every path from x to y must contain v. So x and y cannot lie on a common cycle; a contradiction to the hypothesis. $\square$

Corollary 6.4.5. *A graph is 2-connected if and only if every pair of edges lie on a common cycle.*

Proof. Suppose G is 2-connected and let e_1 and e_2 be any pair of edges. Subdivide e_1 and e_2 by placing vertices v_1 and v_2, respectively, on them to form paths of length 2 and call this subdivision G'. Then G' remains 2-connected. Theorem 6.4.4 implies that v_1 and v_2 lie on a common cycle in G', and therefore e_1 and e_2 lie on a common cycle in G. The converse follows immediately from Theorem 6.4.4. $\square$

Let G be a graph and H be an induced subgraph of G and consider $G\backslash H$, leaving intact isolated vertices that may be forced when removing the edges of H. A *path addition* to H (also called H-*path*) is a path in $G\backslash H$ connecting two distinct vertices of H. If the two vertices of H are not distinct, then it is called a *cycle addition* (or H-*cycle*). Figure 6.10 displays a path addition and a cycle addition. The next two results describe the structure of 2-connected graphs (Whitney, 1932b) and 2-edge connected graphs (Robbins, 1939).[4]

Theorem 6.4.6. (Open Ear Decomposition) *A graph is 2-connected if and only if it can be constructed from a cycle by successively adding path additions to graphs already constructed.*

Proof. Suppose G is 2-connected. Then every pair of vertices lie on a cycle by Theorem 6.4.4. If G itself is a cycle, then there is nothing to show. Therefore assume G is not a cycle. Suppose, if possible, the result fails for G, and let H be the largest proper subgraph of G constructed from a cycle by successively adding path

---

4 Path additions and cycle additions are called *open ears* and *ears*, respectively. The constructions described in the statements of Theorems 6.4.6 and 6.4.7 are called open ear decompositions and ear decompositions, respectively. An *ear decomposition* of G is a partition of the edge set into a sequence of ears $C_0, C_1, \ldots, C_t$, where the first ear C_0 is a cycle and $C_1, \ldots, C_t$ are path or cycle additions. An open ear decomposition is formed by path additions and an ear decomposition is formed by path and cycle additions. Open ear decompositions and ear decompositions are also called *Whitney Synthesis* and *Whitney-Robbins Synthesis*, respectively, in (Gross and Yellen, 2005).

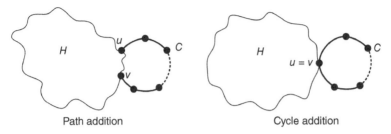

Path addition — Cycle addition

Figure 6.10   Ear decompositions.

additions to graphs already constructed. Observe that H is an induced subgraph since any edge xy in $E(G) - E(H)$ with x, y in $V(H)$ defines a path addition. Since $H \neq G$, there is an edge $e = uv$ such that $u \in V(H)$ and $v \in V(G \backslash H)$. Moreover, since H is an induced subgraph and G is 2-connected, $G - u$ contains a path P with edges outside H from v to a vertex w in H, where $w \neq u$. Edge uv together with P forms a path addition to H; a contradiction to the maximality of H. The converse follows immediately from Theorem 6.4.4, since if G is constructed from a cycle by successively adding path additions, then every pair of vertices lies on a cycle.   □

The last result of this section, stated without proof, is similar to Theorem 6.4.6, except that cycle additions are allowed.

Theorem 6.4.7. (Ear Decomposition)   *A graph is 2-edge-connected if and only if it can be constructed from a cycle by successively adding path additions and cycle additions to graphs already constructed.*

6.5   Menger's Theorem

Let G be a graph and u and v be a pair of distinct vertices (possibly adjacent). Two paths from u to v are called *internally disjoint $u - v$ paths* if they have no common vertices except u and v. Let $\lambda(u, v)$ be the maximum number of pairwise internally disjoint $u - v$ paths.

Let u and v be a pair of non-adjacent vertices. A $u - v$ *separator* is a set of vertices whose removal leaves no $u - v$ path in G. Let $\kappa(u, v)$ be the size of the smallest $u - v$ separator. This measure $\kappa(u, v)$ may be viewed as a measure of "local" connectivity between vertices u and v. Observe that

$$\kappa(u, v) \geq \kappa(G)$$

since $\kappa(G)$ is the size of the smallest vertex-cut. Further observe that

$$\kappa(u, v) \geq \lambda(u, v)$$

since at least one vertex distinct from u and v must be removed from each of the $\lambda(u, v)$ internally disjoint paths to disconnect u and v, as shown in the following diagram:

Menger's theorem establishes that $\kappa(u, v) = \lambda(u, v)$. In other words, the minimum number of vertices whose removal disconnects u and v is equal to the maximum number of internally disjoint $u - v$ paths (Menger, 1927). If there are no edges joining the pairwise internally disjoint $u - v$ paths, then clearly equality holds, as shown in the diagram earlier. The trouble is there may be edges between vertices of the pairwise internally disjoint $u - v$ paths as shown in Figure 6.11. Then it is not so easy to see why equality must hold. In this case, the three internally disjoint paths between u and v are: uav, $ubfv$, and $udecv$.

Figure 6.11 An example for Menger's Theorem

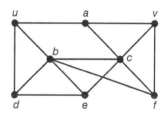

Theorem 6.5.1. (Menger's Theorem) *Let G be a connected graph and let u and v be non-adjacent vertices. Then $\kappa(u, v) = \lambda(u, v)$.*

We select for presentation one of Diestal's proofs of a slightly stronger statement (Diestel, 2017). Instead of two non-adjacent vertices u and v, we will consider two subsets of vertices A and B. An $A - B$ *path* is a path having first vertex in A and last vertex in B and no other vertex in $A \cup B$. An $A - B$ separator is a set of vertices whose removal leaves no $A - B$ path in G. The size of the smallest $A - B$ separator is denoted by $\kappa(A, B)$. Two $A - B$ paths are called *internally disjoint A − B paths* if they have no common vertex except possibly their end vertices. Let $\lambda(A, B)$ be the maximum number of pairwise internally disjoint $A - B$ paths.

Theorem 6.5.2. *Let G be a connected graph and let A and B be subsets of vertices. Then $\kappa(A, B) = \lambda(A, B)$.*

Proof. Clearly $\kappa(A, B) \geq \lambda(A, B)$ since at least one vertex must be removed from each of the $\lambda(A, B)$ pairwise internally disjoint paths. We will prove that $\kappa(A, B) = \lambda(A, B)$ by induction on the number of edges. If G has no edge, then the result is trivially true, so assume that G has at least one edge $e = uv$.

Suppose, if possible, $\lambda(A, B) < \kappa(A, B)$. Consider G/e where, if one of the end vertices of $e = uv$ is in A or B, then the contracted vertex v_e formed in G/e will be in A or B, respectively. By induction

$$\kappa_{G/e}(A, B) = \lambda_{G/e}(A, B) \leq \lambda(A, B) < \kappa(A, B).$$

Let Y be an $A - B$ separator in G/e such that

$$|Y| = \kappa_{G/e}(A, B) < \kappa(A, B).$$

Then $v_e \in Y$, since otherwise Y would be an $A - B$ separator in G of size less than $\kappa(A, B)$. Let

$$X = (Y - \{v_e\}) \cup \{u, v\}.$$

Then X is an $A - B$ separator in G and $|X| = \kappa(A, B)$.

Next, consider $G \backslash e$. Since $u, v \in X$, every $A - X$ separator in $G \backslash e$ is also an $A - B$ separator in G, and therefore has at least $\kappa(A, B)$ vertices. Similarly, every $X - B$ separator in $G \backslash e$ is also an $A - B$ separator in G with at least $\kappa(A, B)$ vertices. By induction, $G \backslash e$ has at least $\kappa(A, B)$ internally disjoint $A - X$ paths and at least $\kappa(A, B)$ internally disjoint $X - B$ paths. Since X is an $A - B$ separator, these paths meet in X and they can be combined to form at least $\kappa(A, B)$ internally disjoint $A - B$ paths. Therefore $\lambda(A, B) = \kappa(A, B)$; a contradiction to our assumption. □

The proof of Theorem 6.5.1 follows from Theorem 6.5.2 by letting $A = \{u\}$ and $B = \{v\}$. Theorem 6.5.1 is a powerful result from which several corollaries follow quite easily.

Recall Theorem 6.4.4, which states that a graph is 2-connected if and only if every pair of vertices lies on a common cycle. In other words a graph is 2-connected if and only if there are two internally disjoint paths between every pair of non-adjacent vertices u and v. Theorem 6.4.4 can be extended to k-connected graphs (Whitney, 1932a).

Proposition 6.5.3. *A graph is k-connected if and only if there are k pairwise internally disjoint paths between every pair of distinct vertices u and v.*

Proof. Suppose G is k-connected and let u and v be a pair of distinct vertices. Then $\kappa(u, v) \geq k$. If u and v are non-adjacent vertices, then Theorem 6.5.1 implies that

$$\lambda(u, v) = \kappa(u, v) \geq k.$$

Therefore we may assume that $e = uv$ is an edge. Suppose, if possible, $\lambda(u, v) \leq k - 1$. Then

$$\lambda_{G\backslash e}(u, v) \leq k - 2$$

and Theorem 6.5.1 implies that

$$\kappa_{G\backslash e}(u, v) = \lambda_{G\backslash e}(u, v) \leq k - 2.$$

Let S be a $u - v$ separator in $G\backslash e$ of size at most $k - 2$. Then either $S \cup \{u\}$ or $S \cup \{v\}$ is a vertex-cut of G. However, both these sets have size at most $k - 1$; a contradiction since G is k-connected. So, $\lambda(u, v) \geq k$.

Conversely, suppose there are k internally disjoint paths between every pair of distinct vertices u and v. Then at least k distinct edges, one along each path, must be removed to disconnect u and v. Since this is true for every pair of vertices, no fewer than k vertices must be removed to disconnect G. Therefore G is k-connected. $\square$

Proposition 6.5.4. *If G is a 3-connected graph, then G has a K_4-minor.*

Proof. Let u and v be two vertices in G. Since G is 3-connected, Theorem 6.5.3 implies that there are three pairwise internally disjoint $u - v$ paths P_1, P_2, and P_3 as shown in the following diagram. Since one of these three paths may be an edge, we may assume two of the three paths have vertices distinct from u and v. Suppose P_1 and P_2 have vertices x and y distinct from u and v. Let Q be a shortest path from x to y not containing u and v. Such a path must exist since G is 3-connected. Additionally, $G - \{u, v\}$ is connected since G is 3-connected. Observe that $P_1 \cup P_2 \cup P_3 \cup Q$ has a contraction-minor isomorphic to K_4 with vertices u, v, x, y. $\square$

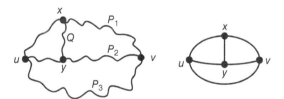

Recall from Section 1.3 that a series-parallel network is a multigraph obtained from a loop or a single edge by subdividing edges or adding edges in parallel. Proposition 6.5.4 leads to an excluded-minor result.

Corollary 6.5.5. *Let G be a 2-connected graph. Then G is a series-parallel network if and only if it has no minor isomorphic to K_4.*

Proof. Observe that K_4 is not a series-parallel network and it is the smallest such graph. Suppose G has no K_4-minor. Then G is not 3-connected by Proposition 6.5.4. Since G is 2-connected, Theorem 6.4.6 implies that G can be constructed from a cycle by successively adding path additions to graphs already constructed. Observe that the only way to avoid a K_4-minor is to add path additions to adjacent vertices. By definition, the graph constructed in this manner is a series-parallel network. □

Recall from Section 1.3 that if G_1 and G_2 are graphs and each has a clique K_t, where $t \geq 2$, then the *t-sum* of G_1 and G_2, denoted by $G_1 \oplus_t G_2$, is formed by identifying the vertices and edges of the clique in both G_1 and G_2 and deleting all the edges of the clique. In the next result we prove that a graph that is 2-connected, but not 3-connected can be decomposed into 2-sums of its 2-connected proper minors.

Theorem 6.5.6. *Let G be a 2-connected graph with at least 6 edges. Then $\kappa(G) = 2$ if and only if $G = H_1 \oplus_2 H_2$, where each of H_1 and H_2 is a 2-connected minor with at least 3 edges.*

Proof. Suppose $\kappa(G) = 2$. Then there is a vertex-cut $\{u, v\}$ whose removal disconnects G. Let x and y be in two distinct components of $G - \{u, v\}$, say $x \in T$ and $y \in (G - \{u, v\}) - T$. First, suppose u and v are non-adjacent vertices. Since $\kappa(G) = 2$, Proposition 6.5.3 implies that in G there are two internally disjoint $x - y$ paths P and Q passing through u and v, respectively, forming a cycle C, as shown in the following diagram.

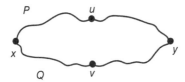

Construct H_1 by deleting all the edges in T outside cycle C and contracting the portion of cycle C in T down to an edge e_1 incident with u and v.

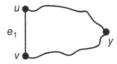

Similarly, construct H_2 by deleting all the edges in $(G - \{x, y\}) - T$ outside cycle C and contracting the portion of cycle C in $(G - \{x, y\}) - T$ down to an edge e_2 incident with u and v. Observe that each of H_1 and H_2 is a 2-connected proper minor of G with at least 3 edges and G is the 2-sum of H_1 and H_2 with edges e_1 and e_2 identified and removed.

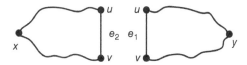

Next, suppose $e = uv$ is an edge. Then include edge e in both H_1 and H_2 to obtain H_1' and H_2', where edge e is parallel to e_1 in H_1 and to e_2 in H_2. Once again G is the 2-sum of proper minors H_1' and H_2' with edges e_1 and e_2 identified and removed.

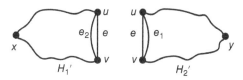

Conversely, suppose $G = H_1 \oplus_2 H_2$. By the definition of 2-sum, there exists an edge e incident with vertices u and v such that $V(H_1) \cap V(H_2) = \{u, v\}$ and $E(H_1) \cap E(H_2) = \{e\}$ and $e = uv$ is removed when forming $H_1 \oplus_2 H_2$. Let $x \in V(H_1) - \{u, v\}$ and $y \in V(H_2) - \{u, v\}$. Every path from x to y must pass through either u or v. So $\{u, v\}$ is a vertex-cut of size 2 in G, and therefore $\kappa(G) = 2$.   □

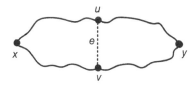

Thus a graph that is not 3-connected can be decomposed into clique sums of its blocks (bridges and 2-connected subgraphs) and by Theorem 6.5.6 the blocks can be decomposed into 2-sums of its 2-connected proper minors. By repeatedly applying Theorem 6.5.6 until no pairs of cut vertices remain, we may conclude that the 2-connected blocks can be decomposed into 2-sums of its 3-connected proper minors. Thus for purposes of understanding the structure of graphs in a particular class, it is enough to focus on the 3-connected graphs in the class.

We will end this chapter by stating without proof two significant structural results for 3-connected graphs. The first is a result by William Tutte called the Wheels Theorem. Some explanation is needed before stating this result. Consider the family of wheel graphs W_{n-1}, where $n \geq 4$. A wheel has two kinds of edges:

Single-edge deletions   Single-edge contractions

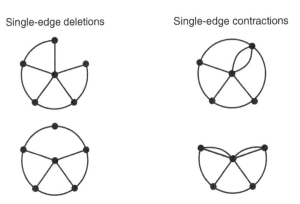

Figure 6.12   Deletion and contraction of the edges of W_5.

rim edges and spoke edges. Now consider the graphs obtained by deleting and contracting rims and spokes. As shown in Figure 6.12, deleting a rim or spoke results in a graph that is 2-connected, but not 3-connected. Contracting a rim or spoke results in a graph that is 3-connected, but not simple due to the presence of the parallel edges. Tutte (1961) proved that this is a special property enjoyed only by wheel graphs. In every other 3-connected graph, there is an edge e such that either $G \backslash e$ or G/e is simple and 3-connected.

Theorem 6.5.7. (Tutte's Wheels Theorem)   *Let G be a simple graph that is not a wheel. Then G is 3-connected if and only if there is an edge e such that either $G \backslash e$ or G/e is simple and 3-connected.*

Seymour (1980) extended Tutte's result by showing that with few exceptions, if G and H are 3-connected graphs such that G has a proper H-minor, then there exists an edge $e \in E(G) - E(H)$ such that either $G \backslash e$ or G/e is simple, 3-connected, and has an H-minor. This is known as the Splitter Theorem.[5] We will end the chapter with this result.

---

5 The result in Seymour (1980) is presented for matroids. Negami (1982) gave a proof for the graph theory version. The hypothesis in the original version states that "if H is a wheel graph then G has no larger wheel". Subsequently, the hypothesis was simplified to only require that G is not a wheel and $H \not\cong W_3$ (Coullard and Oxley, 1992).

Theorem 6.5.8. (Seymour's Splitter Theorem) *Suppose G and H are simple 3-connected graphs such that G has a proper H-minor, G is not a wheel, and H $\not\cong W_3$. Then there is an edge $e \in E(G) - E(H)$ such that either G\e or G/e is simple and 3-connected with an H-minor.*

Each topic in this chapter merits an entire book, and what is presented here is just a taste of the beautiful results in graph structure theory.

Exercises

6.1 For the graphs in Figures 2.8, 3.3, and 3.4:
   (a) Find, if possible, an Eulerian circuit, a Hamiltonian cycle, a cycle decomposition, and a cycle double cover.
   (b) Solve the Postman Problem by determining the least number of edges required to Eulerize the graph. Display the postman's route.
   (c) Find the chromatic number, clique number, vertex connectivity, and edge connectivity.
   (d) The number of colors required by Algorithm 6.3.3 for the worst coloring of a graph is called its *greedy number*. Find the greedy number of the graphs.

6.2 Let G be a connected graph and let C be a cycle and D be a minimal edge-cut. Show that $|C \cap D|$ is even.

6.3 Prove that for any edge e in G, $\lambda(G) - 1 \leq \lambda(G - e) \leq \lambda(G)$.

6.4 Prove that for any vertex v in G, $\kappa(G) - 1 \leq \kappa(G - v)$. Give an example to show that $\kappa(G - v)$ is not necessarily bounded by $\kappa(G)$.

6.5 Let G be a strongly connected digraph with $m \geq 3$ arcs. A *directed Eulerian circuit* is a directed circuit that contains every arc. A *directed Hamiltonian cycle* is a directed cycle that contains every vertex.
   (a) Prove that G has a directed Eulerian circuit if and only if $indeg(v) = outdeg(v)$ for every vertex v.
   (b) Prove that G has a directed Eulerian trail from a to b if and only if $indeg(a) = outdeg(a) - 1$, $indeg(b) = outdeg(b) + 1$, and $indeg(v) = outdeg(v)$ for every vertex $v \neq a, b$.
   (c) Prove that if $deg(u) + deg(v) \geq 2n - 1$ for every pair of non-adjacent vertices u and v, then G has a directed Hamiltonian cycle.
   (d) For the digraphs in Figures 2.8 and 3.3 find, if possible, a directed Eulerian circuit and a directed Hamiltonian cycle.

6.6 Show that the Petersen graph satisfies Theorem 6.2.2, but has no Hamiltonian cycle.

6.7 Prove that a series-parallel network is 3-colorable.

6.8 Prove that for $k \geq 2$, if G is k-connected, then every set of k vertices lie on
a common cycle. Give an example to show that the converse is not true
(Dirac, 1960).

6.9 Let G be a graph and u and v be a pair of distinct vertices (possibly adjacent).
Two paths from u to v are called *edge-disjoint $u - v$* paths if they have no
common edges.
(a) Prove that the maximum number of edge-disjoint $u - v$ paths equals
the minimum number of edges whose removal disconnects u and v.
(b) Prove that G is k-edge-connected if and only if for every pair of distinct
vertices u and v there exist k edge-disjoint $u - v$ paths.

Topics for Deeper Study

6.10 Let G be a connected graph with $n \geq 3$ vertices and degree sequence $d_1 \leq
d_2 \leq \cdots \leq d_n$. Prove that, if there is no integer $k < \frac{n}{2}$ for which $d_k \leq k$ and
$d_{n-k} \leq n - k - 1$, then G is Hamiltonian (Chvátal, 1972).

6.11 Prove that if G is a connected graph, then $\chi(G) \leq max\{\delta(H)\} + 1$, where
the maximum is taken over all induced subgraphs H of G (Szekeres and
Wilf, 1968).

6.12 Let G be a graph with $n \geq 2$ vertices. Prove that if $\delta(G) \geq \frac{n}{2}$, then $\lambda(G) =
\delta(G)$. Give an example to show that the converse is not true (Chartrand,
1966).

6.13 Let G be a graph with $n \geq 2$ vertices and let $1 \leq t \leq n - 1$. Prove that if
$\delta(G) \geq \left\lceil \frac{n+t-2}{2} \right\rceil$, then $\kappa(G) = t$. Give an example to show that the converse
is not true (Chartrand and Harary, 1968).

6.14 Prove Theorem 6.3.5 (Vizing's Edge Coloring Theorem), Theorem 6.3.6
(Lovász' Perfect Graph Theorem), Theorems 6.5.7 (Tutle's Wheels
Theorem) and 6.5.8 (Sermour's Splitter Theorem).

6.15 A graph is said to have a *strongly connected orientation* if it can be converted
to a digraph that is strongly connected. Prove that a connected graph G has

a strongly connected orientation if and only if G has no (undirected) bridges (Robbins, 1939). This result was motivated by a traffic problem. Assuming all the streets in a town are two-way streets, traffic-flow may be modeled as a connected graph with streets as edges and intersections of streets as vertices. When traffic is heavy some two-way streets are turned into one-way streets to reduce congestion. Turning a graph into a digraph is called giving the graph an "orientation." In order to drive from any place to any other place, the resulting digraph must be strongly connected, and as the result states this is possible if and only if there are no bridges.

6.16 For $k \geq 2$, a graph is *critically k-connected* if $\kappa(G) = k$ and $\kappa(G - v) = k - 1$ for every vertex v of G. Prove that if G is critically k-connected, then $\delta(G) < \frac{3k-1}{2}$ (Chartrand et al., 1972).

6.17 For $k \geq 2$, a graph is *minimally k-connected* if $\kappa(G) = k$ and $\kappa(G\backslash e) = k - 1$ for every edge e of G. The following results by Halin and Mader may be found in Bollobás (2004).
 (a) Prove that $\delta(G) = k$.
 (b) Prove that G has at least $\frac{(k-1)n+2}{2k-1}$ vertices of degree k.
 (c) Prove that G is $(k + 1)$-colorable.
 (d) Prove that if $n \geq 3k - 2$, then $m \leq k(n - k)$. Moreover, if $n \geq 3k - 1$, equality holds precisely for $G \cong K_{k,n-k}$.

6.18 Mycielski gave a method for constructing an infinite family of graphs without triangles with chromatic number as high as desired. Let G be a triangle-free graph with n vertices $v_1, \ldots, v_n$ and m edges. Construct G' from G recursively as follows: add n vertices $u_1, \ldots, u_n$ and if $v_i v_j$ is an edge, then make $u_i v_j$ and $u_j v_i$ edges, and add an additional vertex w and join w to vertices $u_1, \ldots, u_n$ (Mycielski, 1955).
 (a) Show that G' is a triangle-free graph with $2n + 1$ vertices, $3m + n$ edges, and $\chi(G') = \chi(G) + 1$.
 (b) Let $G = K_2$ and compute G' and G'' (called the Grötzsch graph). Show that G'' is the smallest 3-connected triangle-free graph with $\chi(G) = 4$ (Chvátal, 1974).

6.19 Theorem 6.3.5 implies there are only two types of graphs with respect to edge coloring: graphs with edge chromatic number $\Delta(G)$ and graphs with edge chromatic number $\Delta(G) + 1$. For example, a cubic graph (regular graph of degree 3) may have edge chromatic number 3 or 4. Thus began the search for connected bridgeless cubic graphs with edge chromatic number 4. Such graphs are called *snarks*.[6] Show that the smallest snark is the Petersen graph. Review the known snarks and the connection between snarks and the Cycle Double Cover Conjecture.

6.20 Let $G \otimes H$ be the tensor product of graphs G and H. Stephen Hedetniemi conjectured in 1966 that $\chi(G \otimes H) = min\{\chi(G), \chi(H)\}$. This became known as Hedetniemi's Conjecture. Yaroslav Shitov showed that this conjecture is not true (Shitov, 2019). Review his paper to understand the counterexample he found.

---

6 These graphs were named snarks by Martin Gardner after the elusive and undefined character named the snark in Lewis Carroll's poem "The Hunting of the Snark" (Gardner, 1976). See also "The Continuing saga of snarks" (Belcastro, 2012).

7

Planar Graphs

Earlier, in Section 1.1, we described a graph as planar if it can be drawn on the plane without crossing edges and non-planar otherwise. We saw that K_5 and $K_{3,3}$ are non-planar, and we reached this conclusion intuitively after making a few attempts to draw them without crossing edges. In this chapter we will undertake a systematic study of planarity. Section 7.1 gathers together several properties of planar graphs, including the fact that every planar graph has a companion graph called its geometric dual. Section 7.2 is on Euclid's famous theorem that there are exactly five Platonic solids. Section 7.3 discusses the Four Color Theorem and presents a proof of the Five Color Theorem. Section 7.4 presents three invariants for non-planar graphs: crossing number, thickness, and genus.

7.1   Properties of Planar Graphs

A drawing of a planar graph on the plane with no edges crossing is called an *embedding* in the plane. The actual drawing is called a *plane representation* of the graph. For example, the graph $K_5 \backslash e$ shown in Figure 1.12 is a planar graph. The second drawing in the figure shows a plane representation of $K_5 \backslash e$. It is understood that a planar graph has a plane representation.

Intuitively, when a planar graph is drawn without crossing edges, the vertices and edges divide the plane into pieces called faces. To make this definition precise we need the concept of an open set and a region, which we will define informally as in West (2001). An open set in the plane is a set O such that for every point $p \in O$, all points within some small distance from p are in O. A region is an open set O that contains a polygonal curve joining every pair of points in O. A *face* of a planar graph embedded in the plane is a maximal region of the plane that contains no point used in the embedding. Each face is surrounded by a cycle called the

Graphs and Networks, First Edition. S. R. Kingan.
© 2022 John Wiley & Sons, Inc. Published 2022 by John Wiley & Sons, Inc.

boundary of the face. Results on planarity require the Jordan Curve Theorem from topology.[1] A simple closed curve is a continuous loop that does not intersect itself.

Theorem 7.1.1. (Jordan Curve Theorem) *A simple closed curve in the plane separates the plane into two disjoint regions with the curve as the boundary of each region.*

Using Theorem 7.1.1 we may assume that any continuous line drawn from the inside of a face to the outside intersects the boundary at a point. For example, the graph in Figure 1.12 has six faces including the outer face (the area on the outside of the graph). This outer face is called the *infinite face* and it must be counted in the number of faces. There is nothing special about the infinite face in a planar graph because any face can be drawn so that it is the infinite face. This is not an obvious statement and requires an understanding of stereographic projection,[2] which is a way of mapping the "punctured sphere" onto the plane. Place the plane tangential to the sphere and consider the point of contact the south pole S as shown in Figure 7.1. The point on the sphere where the line perpendicular to the plane intersects it is the north pole N. The sphere with point N removed is called a *punctured sphere*. Consider a point P on the plane and draw a straight line between N and P. Mark the point where segment NP intersects the sphere as P'. In this manner we can map each point on the plane to a unique point on the punctured sphere. Conversely, consider a point P' in the punctured sphere and draw a straight line between N and P'. Extend segment NP' so that it intersects the plane at P. Thus there is a one-to-one correspondence between points on the punctured sphere and points on the plane. So a graph can be embedded on the surface of a sphere if and only if it can be embedded in the plane. The infinite face on the plane

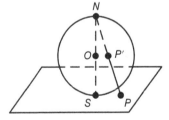

Figure 7.1 Stereographic projection.

---

1 A proof of the Jordan Curve Theorem may be found in Courant et al. (1996). Chapter 5 of their book "What is Mathematics" is titled "Topology," but is mostly graph theory. The field has come a long ways today with applications in every discipline. Indeed, it was already established in 1969 when Frank Harary dedicated his book as follows "To Kasimir Kuratowski who gave K_5 and $K_{3,3}$. To those who thought planarity was nothing but topology."
2 François d'Aiguillon gave stereographic projection its name, although it was known before by cartographers. The term appeared in his 1613 work *Six Books of Optics*.

is mapped to the region around the north pole N. Since on the sphere any point can be designated as the north pole, any face in the plane can become the infinite face. We may conclude that a planar graph can be embedded in a plane so that any face can be made the infinite face (Deo, 1974).

Vertices, edges, and faces are reminiscent of polyhedra from Euclidean geometry. Sixteen years after publishing the first paper on graph theory Euler proved a formula relating the number of vertices, edges, and faces of a planar graph (Euler, 1752).[3] The ancient Egyptians studied polyhedra, as did the ancient Greeks, however, there is no mention of any such formula. Joseph Malkevitch[4] described the importance of Euler's formula as follows:

> Occasionally a mathematical fact is so significant that the results it leads to give insights which range widely over a large area of mathematics and its applications. This is true of Euler's seminal polyhedral formula. Not only did the result lead to deep work in topology, algebraic topology, and the theory of surfaces in the 19th and 20th centuries, but the result has also had great significance for combinatorics, graph theory, computational geometry, and other parts of mathematics. This research continues to be actively pursued today. So much power from just the simple relationship: *vertices + faces − edges = 2.*

Theorem 7.1.2. (Euler's Planar Graph Theorem)   *Let G be a connected planar graph with n vertices, m edges, and f faces. Then $n − m + f = 2$.*

Proof. The proof is by induction on the number of edges $m \geq 0$. The result holds for $m = 0$ since the trivial graph has $n = 1$, $m = 0$, and $f = 1$. Assume the result is true for all connected planar graphs with $m − 1$ edges, and let G be a connected planar graph with m edges. If G is a tree, then $f = 1$ and Proposition 2.1.2 implies that

$$n − m + f = n − (n − 1) + 1 = 2.$$

Therefore suppose G is not a tree. Then G has a cycle and let e be an edge on this cycle. Observe that $G \backslash e$ is a connected planar graph with n vertices, $m − 1$ edges, and $f − 1$ faces since one face disappeared when edge e was removed from the cycle. By the induction hypothesis

$$n − (m − 1) + (f − 1) = 2$$

and therefore

$$n − m + f = 2. \qquad \qquad \square$$

---

3 Euler's paper talked only of polyhedra. Cauchy reformulated Euler's result as a theorem on planar graphs in a sense closest to the modern usage of the term (Cauchy, 1813).
4 See the AMS Feature Column "Euler's Polyhedral formula: Part II" located at http://www.ams .org/publicoutreach/feature-column/fcarc-eulers-formulaii.

A planar graph G is called *maximal planar* if adding one more edge between non-adjacent vertices destroys planarity. Observe that the boundary of every face in a maximal planar graph is a cycle of length 3. The next proposition gives a bound on the number of edges in a planar graph.

Proposition 7.1.3. (Euler's Planar Graph Theorem)   *Let G be a planar graph with $n \geq 3$ vertices and m edges. Then $m \leq 3n - 6$.*

Proof. Let G' be a maximal planar graph on n vertices constructed by adding edges to G. Let m' and f' be the number of edges and faces, respectively, of G'. Then G' is a connected planar graph in which every face is a triangle and every edge is counted in two faces. Therefore

$$2m' = 3f'.$$

Theorem 7.1.2 applied to G' implies that

$$f' = m' - n + 2.$$

Using $f' = \frac{2m'}{3}$, gives

$$\frac{2m'}{3} = m' - n + 2$$
$$2m' = 3(m' - n + 2)$$
$$m' = 3n - 6.$$

Finally, since G has fewer edges than G', $m \leq 3n - 6$.   □

Proposition 7.1.3 gives a way of proving that a graph is non-planar. For example, a planar graph on 5 vertices can have at most 9 edges. Since K_5 has 10 edges, Proposition 7.1.3 implies that K_5 is not planar. It does not, however, prove that $K_{3,3}$ is non-planar. This is because a planar graph on 6 vertices can have at most 12 edges and $K_{3,3}$ has only 9 edges.

The next proposition gives an upper bound on the number of edges of a planar graph in terms of its girth.

Lemma 7.1.4.   *Let G be a bridgeless planar graph with m edges, f faces, and girth $g \geq 3$. Then $2m \geq gf$.*

Proof. Let f_k be the number of faces in G bounded by a cycle of length k, where $k \geq g \geq 3$. Since G is a connected bridgeless planar graph, every edge is counted in two faces. So

$$2m = \sum_{k \geq g} k f_k \geq \sum_{k \geq g} g f_k = g \sum_{k \geq g} f_k = gf.$$

Therefore $2m \geq gf$.   □

Proposition 7.1.5.   *Let G be a bridgeless planar graph with n vertices, m edges, f faces, and girth g $\geq$ 3. Then m $\leq \frac{g}{g-2}(n-2)$.*

Proof. Theorem 7.1.2 implies that $f = m - n + 2$ and Lemma 7.1.4 implies that $\frac{2m}{g} \geq f$. Therefore

$$\frac{2m}{g} \geq m - n + 2$$

and consequently,

$$m \leq \frac{g}{g-2}(n-2).$$

$\square$

Proposition 7.1.5 can be used to show that $K_{3,3}$ is non-planar. The girth of $K_{3,3}$ is 4. For a planar graph with girth $g = 4$,

$$m \leq \frac{g(n-2)}{g-2} = \frac{4(4)}{2} = 8.$$

However, $K_{3,3}$ has 9 edges. Therefore Proposition 7.1.5 implies that $K_{3,3}$ is non-planar.

The next result establishes that a planar graph has at least one vertex of small degree.

Proposition 7.1.6.   *Let G be a planar graph with n $\geq$ 3 vertices. Then $\delta(G) \leq 5$.*

Proof. Suppose all the vertices have degree at least 6. Proposition 1.1.1 implies that

$$2m = \sum deg(v) \geq 6n.$$

So $m \geq 3n$. However, Proposition 7.1.3 implies that $m \leq 3n - 6$. This contradiction completes the proof.

$\square$

As a corollary of Proposition 7.1.6, we can bound the connectivity of planar graphs. The proof is straightforward.

Corollary 7.1.7.   *Let G be a planar graph with n $\geq$ 3 vertices. Then $\kappa(G) \leq 5$.*

The next proposition establishes that a connected planar graph with minimum degree 3 must have at least one face that is bounded by a cycle of length 3, 4, or 5. We begin with a useful lemma.

Lemma 7.1.8. *Let G be a connected planar graph with $\delta(G) \geq 3$. Then $2m \geq 3n$.*

Proof. Let G be a planar graph with n vertices and m edges. Let v_k be the number of vertices of degree k. Then since every vertex has degree at least 3

$$2m = \sum deg(v) = \sum_{k \geq 3} kv_k \geq \sum_{k \geq 3} 3v_k = 3\sum_{k \geq 3} v_k = 3n.$$

Therefore $2m \geq 3n$. □

Proposition 7.1.9. *Let G be a connected bridgeless planar graph with $\delta(G) \geq 3$. Then G has at least one face bounded by a cycle of length 3, 4, or 5.*

Proof. Suppose, if possible, every face of G is bounded by a cycle of length k, where $k \geq 6$. Then girth $g \geq 6$ and by Lemma 7.1.4

$$2m \geq gf \geq 6f.$$

Therefore

$$m \geq 3f.$$

Theorem 7.1.2 implies that $2 = n - m + f$ and Lemma 7.1.8 implies that $n \leq \frac{2m}{3}$, so

$$2 = n - m + f \leq \frac{2m}{3} - m + \frac{m}{3} = 0,$$

which is absurd. Therefore at least one face is bounded by a cycle of length 3, 4, or 5. □

Let G be a planar graph drawn in the plane without crossing edges. The *geometric dual* of G, denoted by G^*, is obtained by putting a vertex in every face and joining two vertices if the corresponding faces have a common edge. Every edge e in G has a corresponding edge e^* in G^*. Figure 7.2 displays $K_5 \setminus e$ and its dual $(K_5 \setminus e)^*$, which is the prism graph from Figure 1.2.

A natural question comes up after drawing the geometric dual of several planar graphs. Is it possible for two non-isomorphic connected planar graphs to have the same duals? The answer is yes, but only in certain cases. Figure 7.3 shows two pairs of planar graphs G_1, G_2 and H_1, H_2. Observe that $G_1 \not\cong G_2$ and $H_1 \not\cong H_2$, but $G_1^* \cong G_2^*$ and $H_1^* \cong H_2^*$. Further observe that each of G_1 and G_2 has a vertex-cut of size 1 and each of H_1 and H_2 has a vertex-cut of size 2. Moreover, G_2 is obtained from G_1 by removing the triangle attached at vertex a and attaching it at vertex b. The graph H_2 is obtained from H_1 by detaching the graph along edge 7 with end vertices a and b, turning the right side upside down and reattaching it along the edge. This is called a "twisting operation." Whitney proved that these are the only two ways that non-isomorphic planar graphs can have isomorphic duals. His key

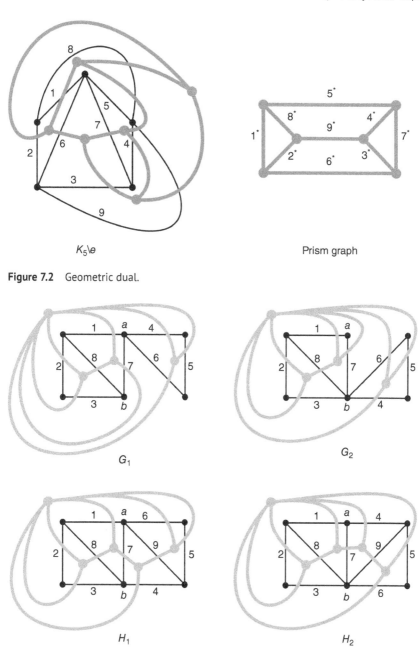

Figure 7.2 Geometric dual.

Figure 7.3 Non-isomorphic graphs with isomorphic geometric duals.

insight was that pairs of graphs such as G_2, G_2 and H_1, H_2 have the same sets of cycles. He proved that 3-connected planar graphs are uniquely determined by their cycles. Consequently, a 3-connected planar graph has a unique dual (Whitney, 1932a, 1933). A proof is available in Oxley (2011).

Theorem 7.1.10. (Whitney's Duality Theorem) *Suppose G_1 and G_2 are 3-connected planar graphs. Then $G_1 \cong G_2$ if and only if $G_1^* \cong G_2^*$.*

We will end this section with Kuratowski's characterization of planar graphs followed by Wagner's interpretation in terms of minors. Together this result is called the Kuratowski-Wagner excluded-minor characterization of planar graphs (Kuratowski, 1930; Wagner, 1937).

Theorem 7.1.11. (Kuratowski's Theorem) *A graph is planar if and only if it has no subgraph that is a subdivision of K_5 or $K_{3,3}$.*

One direction of Theorem 7.1.11 is straightforward. Since K_5 and $K_{3,3}$ are non-planar, any subgraph that is a subdivision of K_5 or $K_{3,3}$ will also be non-planar. Thus if a graph is planar, then it has no subgraph that is a subdivision of K_5 or $K_{3,3}$. The hard work is in the other direction.

Theorem 7.1.12. (Kuratowski-Wagner Theorem) *A graph is planar if and only if it has no minor isomorphic to K_5 or $K_{3,3}$.*

By Theorem 7.1.11 we must prove that G has a subgraph that is a subdivision of K_5 or $K_{3,3}$ if and only if G has a minor isomorphic to K_5 or $K_{3,3}$. Again, one direction is straightforward. Suppose H is a subgraph of G that is a subdivision of K_5. Then H is obtained from K_5 by placing vertices along edges of K_5. So an edge e with a newly added vertex placed on it becomes two edges. To reverse this operation, contract one of the newly formed edges. Thus K_5 is a contraction-minor of H and consequently a minor of G. Similarly, if H is a subgraph of G that is a subdivision of $K_{3,3}$, then $K_{3,3}$ is a contraction-minor of G. The other direction is non-trivial. Proofs of Theorems 7.1.11 and 7.1.12 may be found in West (2001) and in Oxley (2011), respectively.

7.2 Euclid's Theorem on Regular Polyhedra

A *convex polyhedron* is one in which the segment joining any two points of the polyhedron is contained in the interior of the polyhedron or on a face of the polyhedron. The Schlegel diagram of a convex polyhedron is the projection of the polyhedron on the plane made by shining a light just above one of its faces (Schlegel, 1886). Figure 7.4 shows the Schlegel diagram of a cube.

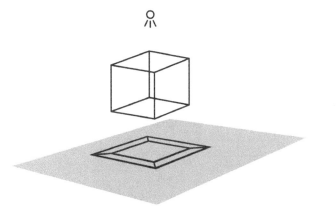

Figure 7.4 Schlegel diagram of a cube.

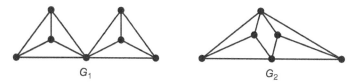

Figure 7.5 Graphs that do not correspond to convex polyhedra.

Not every planar graph corresponds to a convex polyhedron. For example a tree does not correspond to a convex polyhedron. Clearly, all the vertices must have degree at least 3. But is this enough? The answer is no. The graphs in Figure 7.5 do not correspond to convex polyhedra despite having minimum degree 3. The first graph is the 1-sum of two copies of K_4. It has a vertex-cut of size 1. The second graph is the 2-sum of two copies of K_4. It has a vertex-cut of size 2. Grünbaum (2007) describes some of the early confusion surrounding this topic as follows:

> During the nineteenth century, the center of attention concerning poly-hedra switched to (more-or-less explicitly declared) convex ones; we shall later return to the less numerous, but still important considerations of other polyhedra. However, a sense of naive trust in the benevolent nature of mathematical objects persisted ... The first to publicly question this attitude was Ernst Steinitz.

Grunbaum interpreted Steinitz's work (Steinitz, 1922) in graph theoretic language as the following result. See Grünbaum (2003) for the proof.

Theorem 7.2.1. (Steinitz Theorem) *A graph is a convex polyhedron if and only if it is 3-connected and planar.*

The next result collects together the properties of a convex polyhedron.

Proposition 7.2.2.   *Let G be a convex polyhedron with n vertices, m edges, and f faces. Then*

(i) $n - m + f = 2$;

(ii) *At least one vertex has degree 3, 4, or 5;*

(iii) *At least one face is a triangle, square, or pentagon;*

(iv) $2m \geq 3f$; *and*

(v) $2m \geq 3n$

Proof. Theorem 7.2.1 implies that a convex polyhedron is a 3-connected planar graph. Observe that (i) follows from Theorem 7.1.2 and (ii) follows from Proposition 7.1.6. Since G is 3-connected, $\delta(G) \geq 3$, and (iii) follows from Proposition 7.1.9. Lemma 7.1.4 implies that $2m \geq 3f$ and Lemma 7.1.8 implies that $2m \geq 3n$. □

A *regular polygon* is a polygon whose edges have the same length and whose angles have the same measure. A *regular convex polyhedron* is a polyhedron which consists of congruent regular polygons whose vertex angles are all congruent.[5] In particular, a regular convex polyhedron is a regular 3-connected graph. One of the most important results in Euclid's *Elements* states that there are exactly five regular convex polyhedra (also called the Platonic solids because they appear in Plato's work). Figure 7.6 displays the Schlegel diagrams of the five regular convex polyhedra along with the number of vertices, edges, and faces. The proof presented below is taken from Chartrand et al. (2011).

Theorem 7.2.3. (Euclid's Theorem)   *There are exactly five regular convex polyhedra.*

Proof. Let P be a regular convex polyhedron and by Theorem 7.2.1 let G be its associated 3-connected planar graph. Let v_k be the number of vertices of degree k and let f_k be the number of faces bounded by a cycle of length k, where $k \geq 3$. Proposition 7.2.2 (i) implies that $m - n - f = -2$. We may write this statement as

$$-8 = 4m - 4n - 4f = 2m + 2m - 4n - 4f.$$

---

5 Euclid assumed that polyhedra are convex and didn't explicitly mention it. Nor did he mention polyhedral angles when defining regular polyhedra. It is possible to have convex polygons whose faces are congruent polygons, but with a different number of faces meeting at a vertex (and therefore different polyhedral angles). See the AMS Feature Column "Why only Five?" located at http://www.ams.org/samplings/feature-column/fcarc-five-polyhedra. Nevertheless, Euclid had the right ideas on regular polyhedra and a respectable proof.

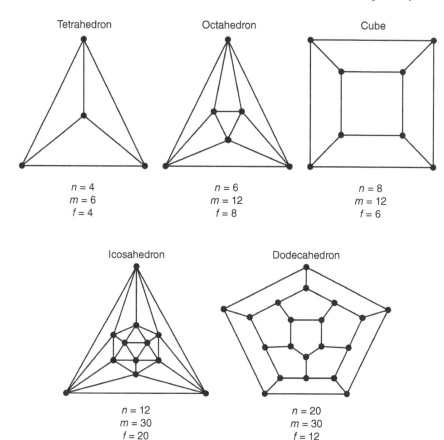

Figure 7.6 Platonic solids.

Therefore

$$-8 = \sum_{k\geq3} kf_k + \sum_{k\geq3} kv_k - 4\sum_{k\geq3} v_k - 4\sum_{k\geq3} f_k = \sum_{k\geq3}(k-4)f_k + \sum_{k\geq3}(k-4)v_k.$$

Since the polyhedron is regular, there exist natural numbers $s \geq 3$ and $t \geq 3$ such that $f = f_s$ and $v = v_t$; that is, $f_k = 0$ for all $k \neq s$ and $v_k = 0$ for all $k \neq t$. Substituting f_s and v_t in the previous equation gives

$$-8 = (s-4)f_s + (t-4)v_t. \tag{7.1}$$

Since each of the s faces of G has f_s edges, $sf_s = 2m$. Likewise, because each of the t vertices of G has degree v_t, $tv_t = 2m$. That is

$$sf_s = tv_t = 2m. \tag{7.2}$$

Moreover, by Proposition 7.2.2

$$3 \leq t \leq 5$$

and

$$3 \leq s \leq 5.$$

Using these bounds on t and s and Eqs. (7.1) and (7.2) we have nine cases:

(i) Suppose $s = 3$ and $t = 3$. Then (7.1) implies that

$$-8 = (3-4)f_3 + (3-4)v_3 = -f_3 - v_3$$

and (7.2) implies that $3f_3 = 3v_3$. Thus we have two equations in two variables

$$\begin{cases} -f_3 - v_3 = -8 \\ f_3 = v_3. \end{cases}$$

Solving these two equations simultaneously gives $f_3 = 4$ and $v_3 = 4$. A regular polygon with $f_3 = 4$ and $v_3 = 4$ is a tetrahedron.

(ii) Suppose $s = 3$ and $t = 4$. Then (7.1) and (7.2) imply that

$$\begin{cases} -8 = -f_3 \\ 3f_3 = 4v_4 \end{cases}.$$

Solving simultaneously gives $f_3 = 8$ and $v_4 = 6$. A regular polygon with $f_3 = 8$ and $v_4 = 6$ is an octahedron.

(iii) Suppose $s = 3$ and $t = 5$. Then (7.1) and (7.2) imply that

$$\begin{cases} -8 = -f_3 + v_5 \\ 3f_3 = 5v_5. \end{cases}$$

Solving simultaneously gives $f_3 = 20$ and $v_5 = 12$. A regular polygon with $f_3 = 20$ and $v_5 = 12$ is an icosahedron.

(iv) Suppose $s = 4$ and $t = 3$. Then (7.1) and (7.2) imply that

$$\begin{cases} -8 = -v_3 \\ 4f_4 = 4v_3 \end{cases}.$$

Solving simultaneously gives $f_4 = 8$ and $v_3 = 6$. A polygon with $f_4 = 6$ and $v_3 = 8$ is a cube.

(v) Suppose $s = 4$ and $t = 4$. Then (7.1) implies $-8 = 0$; a contradiction.

(vi) Suppose $s = 4$ and $t = 5$. Then (7.1) implies $-8 = v_5$; a contradiction.

(vii) Suppose $s = 5$ and $t = 3$. Then (7.1) and (7.2) imply that

$$\begin{cases} -8 = f_5 - v_3 \\ 5f_5 = 3v_3 \end{cases}$$

Solving simultaneously gives $f_5 = 12$ and $v_3 = 20$. A polygon with $f_5 = 12$ and $v_3 = 20$ is a dodecahedron.

(viii) Suppose $s = 5$ and $t = 4$. Then (7.1) implies $-8 = f_5$; a contradiction.

(ix) Suppose $s = 5$ and $t = 5$. Then (7.1) implies $-8 = f_5 + v_5$; a contradiction.

This completes the proof. □

7.3 The Five Color Theorem

In this section we return to the famous Four Color Theorem discussed in Section 1.1. A map is a connected bridgeless planar graph. A *face coloring* of a map is an assignment of colors to the faces so that no two adjacent faces get the same color. Faces that meet in a point may receive the same color. Coloring the faces of a map is equivalent to coloring the vertices of its geometric dual, which is also a connected planar bridgeless graph. Appel and Haken's Four Color Theorem can be stated as a result about the chromatic number.

Theorem 7.3.1. (Four Color Theorem) *Let G be a connected planar bridgeless graph with $n \geq 3$ vertice. Then $\chi(G) \leq 4$.*

The history of the Four Color Theorem is quite interesting and illustrates how a conjecture can generate all sorts of new results. In 1879 Sir Alfred Bray Kempe published a proof of the Four Color Conjecture (Kempe, 1879), but later Heawood found an error in it and published a counterexample (see Sipka (2002)). Nevertheless, he was able to prove the Five Color Theorem (Heawood, 1890).

Theorem 7.3.2. (Five Color Theorem) *Let G be a connected planar bridgeless graph with $n \geq 3$ vertice. Then $\chi(G) \leq 5$.*

Proof. The proof is by induction on the number of vertices. The result holds for $n = 3$ since the only connected planar bridgeless graph with 3 vertices is K_3. Assume the result holds for all connected planar bridgeless graphs with $n - 1$ vertices, and let G be a connected planar bridgeless graph with n vertices.

Theorem 7.1.6 implies that G has a vertex v such that $deg(v) \leq 5$. Consider the subgraph $G - v$, which is planar since G is planar. By the induction hypothesis $\chi(G - v) \leq 5$. Let $\{1, 2, 3, 4, 5\}$ be a 5-coloring of $G - v$. If $deg(v) \leq 4$, then v has four neighbors that use four of the five colors, and the unused color may be assigned to v to obtain a 5-coloring of G. If $deg(v) = 5$ and fewer than five colors are used for the neighbors of v, then again an unused color may be assigned to v to obtain a 5-coloring of G. Therefore suppose $deg(v) = 5$ and the five neighbors of v, $\{v_1, v_2, v_3, v_4, v_5\}$, use colors $1, 2, 3, 4, 5$, respectively.

Without loss of generality select v_1 and v_3, colored 1 and 3, respectively, and consider the Kempe chain H_{13} with respect to the 5-coloring (that is the induced subgraph of G consisting of vertices colored 1 and 3). Since $H_{1,3}$ may be disconnected, v_1 and v_3 may be in the same component or in different components.

First, suppose v_1 and v_3 are in different components of $H_{1,3}$. In this case there is no path from v_1 to v_3 in H_{13}. Recolor v_1 with color 3 and exchange all the colors in the component containing v_1. This frees color 1, which may be assigned to v to obtain a 5-coloring of G.

Next, suppose v_1 and v_3 are in the same component of H_{13}. In this case there is a path P from v_1 to v_3 in $H_{1,3}$, all of whose vertices are colored alternatively 1 and 3. This path P and the path $v_1 v v_3$ produces a cycle in G that surrounds at least one of the remaining vertices, say v_2, as shown in the following diagram.

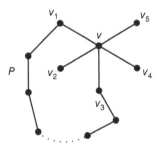

Since G is planar there will be no path from v_2 to v_4, all of whose vertices are colored c_2 or c_4. Thus in the Kempe chain $H_{2,4}$ with respect to the 5-coloring, vertices v_2 and v_4 will be in different components, and the result holds by applying the first case to vertices v_2 and v_4. □

Subsequently, Tait proved that the Four Color conjecture is equivalent to proving that a 3-connected cubic planar graph has a 3-edge coloring. Then he proved that a Hamiltonian 3-connected cubic planar graph has a 3-edge coloring (Tait, 1884). Tait mistakenly assumed that all 3-connected cubic planar graphs are Hamiltonian and concluded that he proved the Four-Color Conjecture. Petersen found the error in Tait's proof (Petersen, 1891) and Tutte found a counterexample to Tait's assertion by displaying a 3-connected cubic planar graph that is not Hamiltonian (Tutte, 1946). This graph is shown in Figure 7.7. The first diagram is called a "Tutte fragment." Observe that if the larger graph in the second diagram contains the fragment, then any Hamiltonian cycle must go in and out of the "top" vertex and one of the lower vertices. It cannot go in and out of the lower vertices. Tutte's graph is formed by placing three fragments with their top vertex pointing inwards, as shown in the third diagram, and joining edges to form a hexagon.

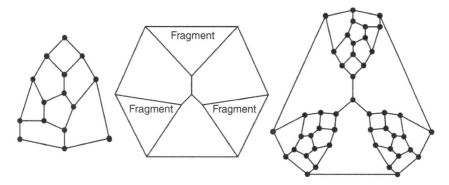

Figure 7.7 Tutte's counterexample to Tait's conjecture.

This graph has 46-vertices, 69 edges, and 25 faces. It is a 3-connected cubic planar non-Hamiltonian graph.[6]

We will end this section by noting that the Four Color Theorem is a special case of Hugo Hadwiger's conjecture (Hadwiger, 1943), which remains unresolved to date.

Conjecture 7.3.3. (Hadwiger's Conjecture) For $t \geq 2$, if G has no K_t-minor, then $\chi(G) \leq t - 1$.

Proposition 1.3.4 implies that G has no K_3-minor if and only if G is a forest, which is bipartite and therefore $\chi(G) = 2$. Corollary 6.5.5 implies that G has no K_4-minor if and only if G is a series-parallel network. It can be shown that if G is a series-parallel network, then $\chi(G) \leq 3$. Wagner proved that the Four Color Theorem is equivalent to proving that if G has no K_5-minor, then $\chi(G) \leq 4$ (Wagner, 1937). More recently, Hadwiger's conjecture was proven for $t = 6$ (Robertson et al., 1993), but it remains open for $t \geq 7$.

7.4 Invariants for Non-planar Graphs

The *crossing number* of a non-planar graph G, denoted by $v(G)$, is the fewest number of crossings necessary to draw the graph in the plane. For example, the crossing number of both K_5 and $K_{3,3}$ is 1 and the crossing number of the Petersen graph is

---

6 See Mulder (1992) for a history of work on the Four Color Problem. Mulder describes Petersen's 1891 paper (Petersen, 1891) as follows: "this paper is remarkable in its depth and scope. It is not just a paper with a new theorem, the nucleus of a new theory was created out of nothing."

2. (We may assume that the crossing number of a planar graph is 0). This invariant was developed by Paul Turán who described the role that crossing numbers played for him during the Second World War as follows[7]:

> In July 1944 the danger of deportation was real in Budapest, and a reality outside Budapest. We worked near Budapest, in a brick factory. There were some kilns where the bricks were made and some open storage yards where the bricks were stored. All the kilns were connected by rail with all the storage yards. The bricks were carried on small wheeled trucks to the storage yards. All we had to do was to put the bricks on the trucks at the kilns, push the trucks to the storage yards, and unload them there. We had a reasonable piece rate for the trucks, and the work itself was not difficult; the trouble was only at the crossings. The trucks generally jumped the rails there, and the bricks fell out of them; in short this caused a lot of trouble and loss of time which was rather precious to all of us (for reasons not to be discussed here). We were all sweating and cursing at such occasions, I too; but nolens-volens the idea occurred to me that this loss of time could have been minimized if the number of crossings of the rails had been minimized. But what is the minimum number of crossings? I realized after several days that the actual situation could have been improved, but the exact solution of the general problem with *m* kilns and *n* storage yards seemed to be very difficult and again I postponed my study of it to times when my fears for my family would end.

Finding $\nu(K_{r,s})$ is now called Turán's Brick Factory Problem, and it is still unsolved. Zarankiewicz (1955) thought he had found the crossing number of $K_{r,s}$, but as it turns out his proof had an error (Guy, 1969). Nevertheless Zarankiewicz established an upper bound and conjectured that there is no better upper bound. In general, finding crossing numbers is an NP-complete problem (Garey and Johnson, 1983).

Conjecture 7.4.1. (Zarankiewicz Conjecture) $\nu(K_{r,s}) = \lfloor \frac{r-1}{2} \rfloor \lfloor \frac{r}{2} \rfloor \lfloor \frac{s-1}{2} \rfloor \lfloor \frac{s}{2} \rfloor$.

The crossing number of K_n is also still unresolved (Guy, 1960; Saaty, 1964).

Conjecture 7.4.2. (Guy-Saaty Conjecture) $\nu(K_n) = \frac{1}{4} \lfloor \frac{n}{2} \rfloor \lfloor \frac{n-1}{2} \rfloor \lfloor \frac{n-2}{2} \rfloor \lfloor \frac{n-3}{2} \rfloor$.

---

7 This paragraph is taken from Turán's 1977 paper welcoming the creation of the *Journal of Graph Theory* (Turán, 1977). See also "The Early History of the Brick Factory Problem" by Beineke and Wilson (2010).

The next result gives a straightforward lower bound on the crossing number in terms of the number of vertices and edges.

Proposition 7.4.3. *Let G be a non-planar graph with n vertices and m edges. Then*

$$v(G) \geq m - 3n + 6$$

Proof. Each crossing in G can be removed by deleting an edge. So the graph formed by deleting $v(G)$ edges is planar. Proposition 7.1.3 implies that

$$m - v(G) \leq 3n - 6$$

and therefore $v(G) \geq m - 3n + 6$. ☐

The lower bound in Proposition 7.4.3 can be used to obtain crossing numbers for some graphs. For example,

$$v(K_6) \geq 15 - 3(6) + 6 = 3.$$

To confirm that $v(K_6)$ is 3 we must display a drawing of K_6 with only 3 crossing edges; a non-trivial task in general, but easy enough for K_6 as shown in Figure 7.8.

A better upper bound for crossing number was found independently by Ajtai et al. (1982) and Leighton (1983). This result, stated without a proof, is called the Crossing Number Inequality.[8]

Theorem 7.4.4. (Crossing Number Inequality) *Let G be a non-planar graph with n vertices and $m \geq 4n$ edges. Then $v(G) \geq \frac{km^3}{n^2}$, where $k \geq 1$ is a constant.*

Figure 7.8 K_6 with 3 edge crossings.

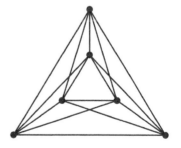

---

8 The proof of the Crossing Number Inequality uses probabilistic methods. Terrance Tao's blog "What's new" has an explanation of the probabilistic method and a proof with $k = \frac{1}{64}$. Additional information is available in "Combinatorial Geometry and its Algorithmic Applications" by Pach and Sharir (2009).

Thickness is another way of gauging the extent of non-planarity. The study of thickness developed with a question in Harary (1961). Consider a graph G on 9 vertices. Is it true that both G and its complement $\overline{G}$ are planar? Since any graph G on 9 vertices has the property that $G \cup \overline{G} = K_9$, if Harary's question is true, then K_9 would be the union of two planar graphs. As it turns out, the answer to Harary's question is no, since K_9 is the union of three planar graphs, not two planar graphs (Battle et al., 1962).

Tutte (1963a) generalized this notion of "biplanarity" and in the process developed the concept of thickness (Tutte, 1963b). The *thickness* of a non-planar graph G, denoted by $\theta(G)$, is the smallest number of pairwise edge-disjoint planar subgraphs whose union is G. For example, the thickness of both K_5 and $K_{3,3}$ is 2. The thickness of a planar graph is 1.

Thickness may be viewed as the smallest number of planes that would be needed to embed G without overlapping edges. One of its applications is to printed circuit-boards. A printed circuit consists of a non-conducting board and a number of terminals. Pairs of terminals must be linked by plating a conducting material on the board. Conductors meet only at terminals, otherwise a short-circuit would occur. Assuming the number of terminals and connections are given, the problem is to locate the terminals and lay out the conductors on the board so that the conductors do not intersect except at their terminals (Foulds et al., 1978; Foulds, 2012). This application is modeled by a planar graph where the vertices represent terminals and conductors joining terminals represent edges. If the graph is non-planar, then the number of insulation layers necessary is the thickness of the corresponding graph. See Mutzel et al. (1998) and Mäkinen et al. (2012) for a survey on thickness.

The next result gives a lower bound on the thickness.

Proposition 7.4.5. *Let G be a non-planar graph with n vertices and m edges. Then*

$$\theta(G) \geq \left\lceil \frac{m}{3n - 6} \right\rceil.$$

Proof. Let G be the union of disjoint planar subgraphs $H_1 \cup H_2 \cup \cdots \cup H_t$, where $t = \theta(G)$ and each H_i has n_i vertices and m_i edges, for $1 \leq i \leq t$. Then Proposition 7.1.3 implies that

$$m = \sum_{i=1}^{t} m_i \leq \sum_{i=1}^{t} \left(3n_i - 6\right) \leq t(3n - 6).$$

Therefore

$$\theta(G) \geq \left\lceil \frac{m}{3n - 6} \right\rceil.$$

$\square$

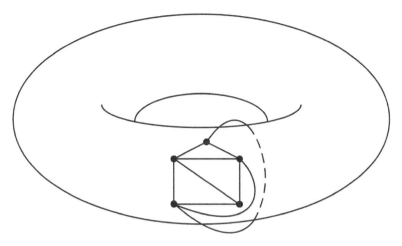

Figure 7.9 K_5 embedded on the torus.

A third way of measuring non-planarity is to ask whether or not a graph can be drawn on other surfaces. For example, K_5 can be drawn or embedded on the torus without crossing edges as shown in Figure 7.9. Every graph can be embedded in a surface with enough holes. To see this, consider any drawing of the graph and attach a "handle" to the plane at each crossing. Allow one edge to go over the handle and the other under it. We can think of the resulting surface as a sphere with handles. The *genus* of a graph, denoted by $\gamma(G)$, is the minimum number of handles that must be added to a sphere so the graph can be embedded in the resulting surface without crossing edges. In this case we can define a face of a non-planar graph in the same way we defined a face in a planar graph. For example, the genus of both K_5 and $K_{3,3}$ is 1. We assume that the genus of a planar graph is 0. In general, for a non-planar graph

$$\gamma(G) \leq \nu(G),$$

since there may be several edges going over and under each handle.

The next result extends Theorem 6.1.1 (Euler's formula). The proof is straightforward, however it does use a major topological result. If two closed surfaces have the same genus, then one may be deformed into the other (Courant et al., 1996).

Theorem 7.4.6. *Let G be a graph with n vertices, m edges, f faces, and genus* γ. *Then*

$$n - m + f = 2 - 2\gamma.$$

Proof. The proof is by induction on the genus γ. Theorem 7.4.2 implies that the result holds for a graph with $\gamma = 0$. Assume that the result holds for graphs with genus $\gamma - 1$. Let G be a graph with genus γ. Let G' be the graph obtained from G by deleting a handle and the k edges that go over it, where $k \geq 1$. Clearly G' has genus $\gamma - 1$, n vertices, and $m - k$ edges. Moreover, G' has $f - k$ faces. To see this, observe that removal of an edge along the handle reduces the number of faces by 1 since the handle connects two distinct faces in G and makes it one face. Thus removal of k edges along the handle reduces the number of faces by k. By the induction hypothesis,

$$n - (m - k) + (f - k) = 2 - 2(\gamma - 1) + 2.$$

Therefore

$$n - m + f = 2 - 2\gamma.$$

$\square$

The term $2 - 2\gamma$ is called the *Euler characteristic* of a surface. The last result in this section gives a lower bound for the genus.

Proposition 7.4.7. *Let G be a non-planar graph with n vertices, m edges, and genus γ. Then*

$$\gamma \geq \left\lceil \frac{m - 3n}{6} \right\rceil + 1.$$

Proof. Let G be embedded in a surface of genus γ so that edges of G do not cross each other. Theorem 7.4.6 states that $n - m + f = 2 - 2\gamma$. By the argument in Lemma 7.1.2, $2m \geq 3f$. Therefore

$$
\begin{aligned}
\gamma &= \frac{2 - n + m - f}{2} \\
&= \frac{6 - 3n + 3m - 3f}{6} \\
&\geq \frac{6 - 3n + 3m - 2m}{6} \\
&= \frac{6 - 3n + m}{6} \\
&= \frac{m - 3n}{6} + 1.
\end{aligned}
$$

Therefore $\gamma \geq \left\lceil \frac{m-3n}{6} \right\rceil + 1.$

$\square$

Exercises

7.1 Show that K_5 and $K_{3,3}$ are the smallest non-planar graphs.

7.2 Prove that if G is a planar graph with n vertices, m edges, f faces, and t components, then $n - m + f = t + 1$.

7.3 Prove that if G is a planar bipartite graph with n vertices and m edges, then $m \leq 2n - 4$.

7.4 Prove that if G is a planar graph, then the average degree $d(G) \leq 6 - \frac{12}{n}$.

7.5 Determine the planar graphs among the graphs in Figures 1.28, 2.8, 3.3, and 3.4 and draw their geometric duals

7.6 Prove that Tutte's graph shown in Figure 7.7 is non-Hamiltonian (Tutte, 1946).

7.7 Determine the crossing numbers, thickness, and genus of K_6, K_7, K_8, K_9, $K_{3,3}$, $K_{3,4}$, $K_{4,4}$, and the Petersen graph.

7.8 Prove that a planar graph can be drawn without crossings so that its edges are straight line segments (István, 1948).

7.9 The *coarseness* of a non-planar graph G, denoted by $\xi(G)$, is the maximum number of edge-disjoint non-planar subgraphs contained in G. Determine the coarseness of K_n and $K_{r,s}$ (Beineke and Chartrand, 1968).

7.10 A graph is *outerplanar* if it can be drawn in the plane so that all of its vertices lie on boundary of the same face. (Since any face can be turned into the outer face, we may assume all of its vertices lie on the boundary of the outer face.)
(a) Determine which of the planar graphs in Figures 1.28, 1.32, 3.3, and 3.4 are outerplanar.
(b) Prove that a graph is outerplanar if and only if it has no minor isomorphic to K_4 or $K_{2,3}$.
(c) Prove that if G is an outerplanar graph with n vertices and m edges, then $m \leq 2n - 3$.

Topics for Deeper Study

7.11 Prove Theorem 7.1.11 (Kuratowski's Theorem) and Theorem 7.1.12 (Wagner's Theorem).

7.12 Prove that the Four Color Problem is equivalent to proving that a 3-connected planar cubic graph has a 3-edge coloring (Tait, 1884).

7.13 Prove that a 3-connected planar cubic Hamiltonian graph has a 3-edge coloring.

7.14 Prove that every 4-connected planar graph is Hamiltonian (Tutte, 1956).

7.15 Counterexamples to Tait's conjecture that "every 3-connected planar cubic graph is Hamiltonian" with fewer vertices than Tutte's graph have been found. Prove that the graph with 44 vertices in Grinberg (1968) and the graph with 42 vertices in Faulkner and Younger (1974) are non-Hamiltonian.

7.16 In 1966 Tutte conjectured that a "2-connected planar cubic graph with no Peterson minor is 3-edge colorable." This became known as "Tutte's Edge Coloring Conjecture." A step toward proving this conjecture was made by Robertson et al. (1997b) and it was resolved by Edwards et al. (2016). Prove that a 2-connected planar cubic graph with no Peterson minor is 3-edge colorable.

7.17 In 1971 Tutte conjectured that "Every 3-connected cubic bipartite graph is Hamitonian" (Tutte, 1971). However multiple counterexamples were found. Prove that the graph with 92 vertices in Horton (1982), the graph with 54 vertices in Ellingham and Horton (1983), and the graph with 50 vertices in Georges (1989) are non-Hamiltonian. Subsequently, David Barnette made a related conjecture that is still open. He conjectured that "Every 3-connected planar cubic bipartite graph is Hamiltonian" (Barnette, 1969).

7.18 Prove that for $r, s \geq 3$, $v\left(K_{r,s}\right) \leq \lfloor \frac{r-1}{2} \rfloor \lfloor \frac{r}{2} \rfloor \lfloor \frac{s-1}{2} \rfloor \lfloor \frac{s}{2} \rfloor$ (Zarankiewicz, 1955);

7.19 Prove that for $n \neq 9, 10$, $\theta(K_n) = \lfloor \frac{n+7}{6} \rfloor$ and $\theta(K_9) = \theta(K_{10}) = 3$ (Alekseev and Gončakov, 1976) and $\theta(K_{t,t}) = \lfloor \frac{t+5}{4} \rfloor$ (Beineke et al., 1964).

7.20 Prove that $\gamma(K_{r,s}) = \lfloor \frac{(r-2)(s-2)}{4} \rfloor$ (Ringel, 1965).

7.21 Prove that $\nu\left(K_{n_1,n_2,n_3}\right) \leq \sum \lfloor \frac{n_j}{2} \rfloor \; \lfloor \frac{n_j-1}{2} \rfloor \; \lfloor \frac{n_k}{2} \rfloor \; \lfloor \frac{n_k-1}{2} \rfloor + \lfloor \frac{n_i}{2} \rfloor \; \lfloor \frac{n_i-1}{2} \rfloor \; \lfloor \frac{n_j n_k}{2} \rfloor$, where the sum is taken over $i = 1, 2, 3$ and $\{j, k\} = \{1, 2, 3\} - i$ (Gethner et al., 2017).

8

Flows and Matchings

Section 8.1 introduces the concept of a flow in a weighted directed graph. The main result in this section is the Max-Flow-Min-Cut theorem (Ford and Fulkerson, 1956). This is the second result where the minimum of one invariant equals the maximum of another invariant. Theorem 6.5.1 (Menger's Theorem) was a min–max theorem. These theorems take different forms, but are equivalent to each other. Section 8.2 introduces four types of sets in graphs and their interactions with each other: stable sets, vertex covers, matchings, and edge covers. Section 8.3 presents additional min–max theorems. Section 8.4 begins with Berge's "augmenting path" theorem (Berge, 1957). The technique used in the proof of this theorem leads to shorter proofs for several earlier results. This section will end with Edmonds "blossom" algorithm for finding a maximum matching. Edmonds (1965a) initiated the study of efficient algorithms with this algorithm and the philosophical discussion in his seminal paper "Paths, Trees and Flowers."

8.1 Flows in Networks

Let G be a weighted directed graph[1] with vertices labeled $1, 2, \ldots, n$. Let $c_{ij} \geq 0$ be the weight on arc (i, j). In this setting we call c_{ij} the *capacity* of arc (i, j) and interpret it as the maximum amount of some commodity that flows through an arc per unit of time in a steady-state situation. The commodity flows through the network like water flows through pipes. Examples of commodities include oil, electricity, messages, people, and trucks. Let x_{ij} be the amount flowing through arc (i, j). We begin by making some assumptions:

(1) The weighted digraph has a starting vertex s called *source* and an ending vertex t called *sink*.

---

1 A weighted directed graph may be called a "weighted digraph" or a "directed network."

Graphs and Networks, First Edition. S. R. Kingan.

(2) The amount flowing through arc (i,j) can be no more than the capacity of the arc. Therefore $0 \leq x_{ij} \leq c_{ij}$. The arc is called *saturated* if $x_{ij} = c_{ij}$.

(3) For all vertices other than the source and the sink, the amount that goes into a vertex must equal the amount that comes out. In other words for vertex $i \neq s, t$

$$\sum_j x_{ij} = \sum_j x_{ji},$$

where the summation is taken over all the neighbors j of vertex i.

(4) The net flow out of the source s must equal the net flow into the sink t.

$$\sum_j x_{sj} - \sum_j x_{js} = \sum_j x_{jt} - \sum_j x_{tj}.$$

This assumption is called the *Conservation Law*.

In summary let G be a weighted digraph with capacity c_{ij} on arc (i,j) and two special vertices called the source s and sink t. An $s-t$ *flow* is a set of non-negative integers x_{ij} assigned to arc (i,j) such that

(i) $0 \leq x_{ij} \leq c_{ij}$;

(ii) $\sum_j x_{ij} = \sum_j x_{ji}$ for all vertices $i \neq s, t$; and

(iii) $\sum_j x_{sj} - \sum_j x_{js} = \sum_j x_{jt} - \sum_j x_{tj} = f.$

The common value f is called the *value* of the flow.

For example, the weighted digraph in Figure 8.1 has capacity 10 on all arcs, to make things simple. The capacity is put in a square box on the first diagram, and assumed to be 10 on the second and third diagram, even though it is not shown. The second diagram has a flow of 6. To see this, observe that the total weight of the edges directed away from s is 6 and the total weight of the edges entering t is also 6.

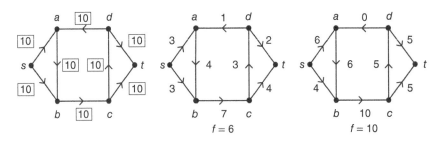

Figure 8.1 Network flows.

Additionally, observe that for each vertex besides s and t, the total weight of edges entering the vertex is the same as the total weight of edges exiting it. Similarly the third diagram in Figure 8.1 has flow 10.

An $s - t$ *cut* is a partition (A, B) of the vertex set such that $s \in A$ and $t \in B$, and the removal of arcs from vertices in A to vertices in B leaves no directed $s - t$ path. Let C be the set of arcs from vertices in A to vertices in B. By a slight misuse of terminology it is customary to call C an $s - t$ cut. The capacity of C is the sum of the capacities of the arcs in C.

The main result in this section is that the largest value of an $s - t$ flow is equal to the smallest capacity of an $s - t$ cut (Ford and Fulkerson, 1956). The authors began their paper with the following quote:

> The problem discussed in this paper was formulated by T. Harris as follows: "Consider a rail network connecting two cities by way of a number of intermediate cities, where each link of the network has a number assigned to it representing its capacity. Assuming a steady state condition, find a maximal flow from one given city to the other."[2]

Theorem 8.1.1. (Max-Flow-Min-Cut Theorem) *Let G be a weighted digraph with source s, sink t, and capacity $c_{ij} \geq 0$ on arc (i, j). The maximum value of an $s - t$ flow equals the minimum capacity of an $s - t$ cut.*

We begin the proof with three lemmas. The hypothesis for all the lemmas is that G is a weighted digraph with source s, sink t, and capacity $c_{ij} \geq 0$ on arc (i, j). In addition, let x_{ij} be an $s - t$ flow, where x_{ij} is an integer assigned to arc (i, j) and $0 \leq x_{ij} \leq c_{ij}$. Let (A, B) be an $s - t$ cut, that is, (A, B) is a partition of the vertex set such that $s \in A$ and $t \in B$. Removal of arcs from vertices in A to vertices in B leaves no directed $s - t$ path. Let C be the set of arcs from vertices in A to vertices in B and let f be the value of an $s - t$ flow.

Lemma 8.1.2. $f = \sum\limits_{i \in A} \sum\limits_{j \in B} (x_{ij} - x_{ji})$.

Proof. Since x_{ij} is an $s - t$ flow,

$$f = \sum_{j} x_{sj} - \sum_{j} x_{js}.$$

---

2 The early history of flow problems and its beginnings in troop transportation during World War II are described in Schrijver (2002).

Note that at each vertex besides s and t, the amount that goes in must come out, so the difference is zero and adding extra terms that are zero does not change the equality. Therefore

$$
\begin{aligned}
f &= \sum_j x_{sj} + \sum_{i \in A, i \neq s} \left[\sum_j x_{ij} - \sum_j x_{ji} \right] - \sum_j x_{js} \\
&= \sum_j x_{sj} + \sum_{i \in A, i \neq s} \sum_j x_{ij} - \sum_{i \in A, i \neq s} \sum_j x_{ji} - \sum_j x_{js} \\
&= \sum_{i \in A} \sum_j x_{ij} - \sum_{i \in A} \sum_j x_{ji} \\
&= \sum_{i \in A} \left[\sum_j x_{ij} - \sum_j x_{ji} \right] \\
&= \sum_{i \in A} \sum_j (x_{ij} - x_{ji}).
\end{aligned}
$$

Since (A, B) is a partition, divide the inner summation $\sum_j (x_{ij} - x_{ji})$ into a sum over $j \in A$ and a sum over $j \in B$ to obtain

$$
\begin{aligned}
f &= \sum_{i \in A} \sum_{j \in A} (x_{ij} - x_{ji}) + \sum_{i \in A} \sum_{j \in B} (x_{ij} - x_{ji}) \\
&= \sum_{i \in A} \sum_{j \in A} x_{ij} - \sum_{i \in A} \sum_{j \in A} x_{ji} + \sum_{i \in A} \sum_{j \in B} (x_{ij} - x_{ji}).
\end{aligned}
$$

Finally, since $\sum_{i \in A} \sum_{j \in A} x_{ij} = \sum_{i \in A} \sum_{j \in A} x_{ji}$, we may conclude that $f = \sum_{i \in A} \sum_{j \in B} (x_{ij} - x_{ji})$. □

The next lemma establishes that the value f of an $s - t$ flow is at most the capacity of an $s - t$ cut.

Lemma 8.1.3. $f \leq \sum_{i \in A} \sum_{j \in B} c_{ij}$

Proof. By Lemma 8.1.2, and since $x_{ij} - x_{ji} \leq x_{ij} \leq c_{ij}$ for arcs with initial vertex in A and terminal vertex in B,

$$
f = \sum_{i \in A} \sum_{j \in B} (x_{ij} - x_{ji}) \leq \sum_{i \in A} \sum_{j \in B} c_{ij}.
$$
□

Let P be an undirected $s - t$ path in the underlying graph. An edge ij in P is called a forward edge if the arrow points from i to j and a backward edge if the arrow points from j to i. The path P is called an $s - t$ flow-augmenting path if $x_{ij} < c_{ij}$ for every forward edge and $x_{ji} > 0$ for every backward edge. For every forward edge ij, let

$$
s_{ij} = c_{ij} - x_{ij}.
$$

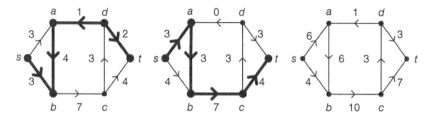

Figure 8.2 Flow augmenting path.

We call s_{ij} the slack in edge ij. Given a specific $s - t$ (undirected) path, consider all the forward and backward edges and take the minimum value over all s_{ij} and x_{ji}. Call this number c and think of it as the capacity of the flow. Observe that $c > 0$ since the flow on forward edges is strictly under capacity and the flow on backward edges is strictly greater than 0. We can increase the overall value of the flow by assigning each backward edge the value $x_{ji} - c$ and each forward edge the value $x_{ij} + c$. An example of this process is shown in Figure 8.2. The weighted digraph has capacity 10 on every edge (not shown to avoid cluttering the figure).

- In the first diagram, the flow-augmenting path highlighted is *sbadt*. Edge *sb* is a forward edge with slack $10 - 3 = 7$. Edges *ba* and *ad* are backward edges with flow 4 and 1, respectively. Edge *dt* is a forward edge with slack $10 - 2 = 8$. The smallest value of the slacks and flows is 1, so the capacity of this flow-augmenting path is $c = 1$. The flow on the forward edges can be increased by 1 and the flow on the backward edges can be decreased by 1.
- In the second diagram, after the changes in the first step are made, the flow-augmenting path *sabct* has capacity $c = 3$ since the four edges *sa*, *ab*, *bc*, and *ct* have slacks 6, 7, 3, and 6, respectively. The flow on the forward edges can be increased by 3 and the flow on the backward edges can be decreased by 3.
- In the third diagram, after the changes in the second step are made, the (undirected) path *sbct* cannot be augmented since forward edge *bc* is at capacity (that is $x_{ij} = c_{ij}$). It is *not* a flow-augmenting path. In fact, any (undirected) $s - t$ path with either a forward edge at capacity or a backward edge with $x_{ji} = 0$ (not necessarily both) cannot be a flow-augmenting path. The process ends at this step.

Note that an $s - t$ flow-augmenting path can be used just once. This is because once its flow is adjusted either the flow along a backward edge will be zero or the slack along a forward edge will be zero. Specifically, if $c = x_{ji}$ where x_{ji} is the flow of a backward edge, then the flow along that backward edge becomes zero. If $c = s_{ij}$ where s_{ij}, is the slack of a forward edge ij, then that forward edge reaches capacity. In the first diagram of Figure 8.1, after the flow is adjusted, the new flow has $c = 0$. So this $s - t$ (undirected) path will no longer admit a flow augmentation. Moreover, augmenting the path, as described, does not create additional flow-augmenting paths. Proceeding in this manner the value of the flow along $s - t$ flow-augmenting paths can be increased until there are no more $s - t$ flow-augmenting paths. The next lemma establishes that the value of the flow is maximum when no $s - t$ flow-augmenting path can be found.

Lemma 8.1.4. *The value of an $s - t$ flow is maximum if and only if there are no $s - t$ flow-augmenting paths.*

Proof. Suppose the value of an $s - t$ flow is maximum. Then there cannot be an $s - t$ flow-augmenting path, since by definition a flow-augmenting path increases the value of the flow. Conversely, suppose there is no $s - t$ flow-augmenting path. Let A consist of all vertices j that have an $s - j$ flow-augmenting path. Clearly $A \neq \phi$ since $s \in A$ and a path of length 0 from s to s is vacuously a flow-augmenting path. Let $B = V(G) - A$. Clearly $t \in B$ since there is no $s - t$ flow augmenting path. Moreover, (A, B) is a partition of the vertex set. For every $i \in A$ and $j \in B$, every forward edge is at capacity $(x_{ij} = c_{ij})$ and every backward edge has $x_{ji} = 0$. Otherwise the edge ij would result in a flow-augmenting path from s to j. Lemma 8.1.2 implies that

$$f = \sum_{i \in A} \sum_{j \in B} (x_{ij} - x_{ji}) = \sum_{i \in A} \sum_{j \in B} (c_{ij} - 0) = \sum_{i \in A} \sum_{j \in B} c_{ij},$$

which is the maximum value of a flow by Lemma 8.1.3. □

We are now ready to prove Theorem 8.1.1, and the proof is quite short since all the work has been done in the lemmas.

Proof of Theorem 8.1.1. Let f be the maximum value of an $s - t$ flow. Lemma 8.1.4 implies that this $s - t$ flow has no flow-augmenting paths and by the construction in the proof of Lemma 8.1.4, there is an $s - t$ cut (A, B) with capacity $\sum_{i \in A} \sum_{j \in B} c_{ij}$ such that

$$f = \sum_{i \in A} \sum_{j \in B} c_{ij}.$$

Thus the capacity of the $s - t$ cut given by the partition (A, B) must be minimum. □

The proof of Theorem 8.1.1 is inherently constructive in nature and gives an algorithm.

Algorithm 8.1.5. Max-Flow-Min-Cut Algorithm

Input: A directed network with capacities c_{ij} and two distinguished vertices s and t.

Output: A maximum (s, t)-flow.

(1) Set $x_{ij} = 0$ for all i, j.
(2) Find an augmenting path P from s to t. If none exists STOP and output x_{ij} as maximum flow.
(3) Find $s_{ij} = c_{ij} - x_{ij}$ for every forward arc of P.
(4) Find c, which is the minimum of s_{ij} for every forward arc in P and x_{ij} for every backward arc in P.
(5) Define a new flow x_{ij} by adding c to all the forward arcs of P and subtracting c from all the backward arcs of P.
(6) Go to Step 2.

This is a high level algorithm and a full analysis of its complexity would require elaboration in Step 2.

Alan Hoffman described Fulkerson's accomplishments in flows as follows:

> His greatest honor is simply that network flows exists as a subject of such importance that all over the world now and in the future, it is and will be a fundamental tool in economic and industrial planning. It was Ray's great good fortune or perhaps the reward of his talent and energy to create mathematics that contribute to life where art and nature imitate each other.

8.2 Stable Sets, Matchings, Coverings

Let G be a graph with n vertices, none of which are isolated. We begin with four invariants associated with a graph.

(1) A subset of vertices, no two of which are adjacent, is called a *stable set* (also called an independent set). The number of vertices in the largest stable set is denoted by $\alpha(G)$.
(2) A subset of edges, no two of which are adjacent, is called a *matching*. The number of edges in the largest matching is denoted by $\beta(G)$.
(3) A subset of vertices incident with every edge is called a *vertex cover*. The number of vertices in the smallest vertex cover is denoted by $\alpha'(G)$.
(4) A subset of edges incident with every vertex is called an *edge cover*. The number of edges in the smallest edge cover is denoted by $\beta'(G)$.

It makes sense to avoid isolated vertices since the presence of isolated vertices artificially increases the largest stable set and makes it impossible to have an edge cover. Moreover, isolated vertices are irrelevant for vertex covers and matchings.

Let us illustrate the definitions with four examples:

(1) For the graph G in Figure 8.3, the set $\{v_1, v_6\}$ is a stable set, but it is not the largest stable set, since $\{v_4, v_6, v_7\}$ is the largest stable set. Hence $\alpha(G) = 3$. The complement of $\{v_4, v_6, v_7\}$ is $\{v_1, v_2, v_3, v_5\}$ and it is a vertex cover. Since there

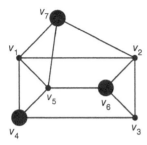 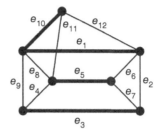

Figure 8.3 Stable sets, matchings, and coverings.

is no vertex cover with fewer vertices, $\alpha'(G) = 4$. The subset of edges $\{e_1, e_3, e_5\}$ is a matching of largest size, hence $\beta(H) = 3$, but it is not an edge cover. The subset $\{e_1, e_3, e_5, e_{10}\}$ is an edge cover. In this graph no set of three edges will serve as an edge cover, so $\beta'(G) = 4$.

(2) For the prism graph G shown in Figure 1.2, two examples of stable sets are $\{v_1, v_6\}$ and $\{v_4, v_6\}$. It is easy to check that there is no stable set with three vertices, so $\alpha(G) = 2$. The complement of $\{v_1, v_6\}$ is $\{v_2, v_3, v_4, v_5\}$ and it is a vertex cover. No set with fewer vertices is a vertex cover, so $\alpha'(G) = 4$. The subset of edges $\{e_1, e_3, e_5\}$ is both a largest size matching and a smallest size edge cover. So $\beta(G) = 3$ and $\beta'(G) = 3$.

(3) Consider K_n, where $n \geq 2$. Observe that $\alpha(K_n) = 1$ since every vertex is adjacent to every other vertex and $\alpha'(K_n) = n - 1$ since $n - 2$ vertices cannot cover all the edges (due to the edge between the remaining two vertices). Moreover, $\beta(K_n) = \lfloor \frac{n}{2} \rfloor$ and $\beta'(K_n) = \lceil \frac{n}{2} \rceil$.

(4) Consider $K_{r,s}$, where $1 \leq r \leq s$. Observe that $\alpha(K_{r,s}) = s$ since selecting the vertices in the larger of the two classes gives the largest stable set and for the opposite reason $\alpha'(K_{r,s}) = r$. Finally, $\beta(K_{r,s}) = r$ and $\beta'(K_{r,s}) = s$.

The next two results relate these parameters to each other.

Proposition 8.2.1. *Let G be a graph with no isolated vertices:*

(i) $\alpha'(G) \geq \beta(G)$; *and*
(ii) $\beta'(G) \geq \alpha(G)$.

Proof. First let X be a vertex cover and M be a matching. A vertex cover must contain at least one vertex incident with every edge in M. So for every edge e in M, there is a vertex v_e in X incident with e. Moreover, $v_e \neq v_f$ for every pair of distinct edges e and f in M since M is a matching. Therefore $|X| \geq |M|$. Since every vertex cover is at least as large as every matching, the smallest vertex cover is at least as large as the largest matching. Consequently $\alpha'(G) \geq \beta(G)$.

Second, let Y be an edge cover and S be a stable set. An edge cover must contain at least one edge incident with every vertex in S. So for every vertex v in S, there is an edge e_v in Y incident with v. Moreover, $e_u \neq e_v$ for every pair of distinct vertices u and v in S since S is a stable set. Therefore $|Y| \geq |S|$. Since every edge cover is at least as large as every stable set, the smallest edge cover is at least as large as the largest stable set. Consequently $\beta'(G) \geq \alpha(G)$. □

Proposition 8.2.2. *Let G be a graph with no isolated vertices. Then S is a stable set if and only if $V - S$ is a vertex cover.*

Proof. Let S be a stable set. Assume that $V - S$ is not a vertex cover. Then there is an edge xy that is not covered by the vertices in $V - S$. So both end vertices x and y must be in S. This is a contradiction to the hypothesis that S a stable set.

Conversely, let $V - S$ be a vertex cover. Assume that S is not a stable set. Then there are vertices $x, y \in S$, such that xy is an edge. So edge xy is not covered by vertices in $V - S$. This is a contradiction to the hypothesis that $V - S$ is a vertex cover. □

A matching that is also an edge cover is called a *perfect matching* (not to be confused with perfect graphs). For example, the graph in Figure 8.3 does not have a perfect matching since there is no matching that is also an edge cover. The prism graph in Figure 1.2 has a perfect matching since $\{e_1, e_3, e_5\}$ is a matching as well as an edge cover. The next result is straightforward.

Proposition 8.2.3. *Let G be a graph with n vertices, where n is even. Then G has a perfect matching if and only if $\beta(G) = \beta'(G) = \frac{n}{2}$.*

Proof. A perfect matching consists of a set of non-adjacent edges that are incident with every vertex. So the size of a perfect matching is $\frac{n}{2}$. Since $\beta(G) \leq \frac{n}{2}$ and $\beta'(G) \leq \frac{n}{2}$, G has a perfect matching if and only if $\beta(G) = \beta'(G) = \frac{n}{2}$. □

In the four examples presented earlier, observe that

$$\alpha(G) + \alpha'(G) = \beta(G) + \beta'(G) = n$$

Tibor Gallai proved that these relations hold for all graphs (Gallai, 1958). His result is the main result of this section. For convenience we call a matching of largest size a "maximum matching." Likewise for other similar expressions.

Theorem 8.2.4. (Gallai's Theorem) *Let G be a graph with n vertices none of which are isolated. Then*

(i) $\alpha(G) + \alpha'(G) = n$
(ii) $\beta(G) + \beta'(G) = n$

Proof. First, let S be a maximum stable set. Then $\alpha(G) = |S|$. Proposition 8.2.2 implies that $V - S$ is a vertex cover. So

$$\alpha'(G) \le |V - S| = |V| - |S| = n - \alpha(G)$$

and therefore

$$\alpha(G) + \alpha'(G) \le n. \tag{8.1}$$

Let W be a minimum vertex cover. Then $\alpha'(G) = |W|$. Proposition 8.2.2 implies that $V - W$ is a stable set. So

$$\alpha(G) \ge |V - W| = |V| - |W| = n - \alpha'(G)$$

and therefore

$$\alpha(G) + \alpha'(G) \ge n. \tag{8.2}$$

Equations (8.1) and (8.2) imply that $\alpha(G) + \alpha'(G) = n$.

Second, let M be a maximum matching. Then $\beta(G) = |M|$. If M is also an edge cover, then M is a perfect matching, and by Proposition 8.2.3 $\beta(G) = \beta'(G) = \frac{n}{2}$ and the result holds. Thus we may assume that M is not an edge cover. Let U be the non-empty set of vertices not covered by M. Then

$$|U| = n - 2|M|.$$

For each vertex $u \in U$ choose an edge incident with u and construct a set M' by adding these $n - 2|M|$ distinct edges to M. Then M' is an edge cover and

$$\beta'(G) \le |M'| = |M| + n - 2|M| = n - |M| = n - \beta(G).$$

Therefore

$$\beta(G) + \beta'(G) \le n. \tag{8.3}$$

Next, let F be a minimum edge cover. Then $|F| = \beta'(G)$. If F contains a cycle, then one edge in the cycle may be removed to obtain a smaller vertex cover; contradicting the minimality of F. Therefore F is a forest. Moreover, since F is an edge cover, all n vertices of G are in F, making F a spanning forest. Let t be the number of connected components in F. Proposition 2.1.3 implies that $|F| = n - t$. Thus $\beta'(G) = n - t$, and consequently $t = n - \beta'(G)$. Construct a matching M by selecting an edge from each of the t components of F. Then

$$\beta(G) \ge |M| = t = n - \beta'(G).$$

Therefore

$$\beta(G) + \beta'(G) \ge n. \tag{8.4}$$

Equations (8.3) and (8.4) imply that $\beta(G) + \beta'(G) = n$. □

Note that if G is a bipartite graph with n vertices, where n is even and none of the vertices are isolated, then Proposition 8.2.3 and Theorem 8.2.4 imply that G is perfect if and only if

$$\beta(G) = \beta'(G) = \alpha(G) = \alpha'(G) = \frac{n}{2}.$$

The aforementioned results are constructive in nature and give methods for computing the four invariants. Proposition 8.2.2 implies that the complement of a stable set is a vertex cover and vice versa. However, this is practical only for small graphs. The proof of Theorem 8.2.4(ii) gives a method for constructing an edge cover from a maximum matching and a matching from a minimum edge cover. Again this is practical only for small graphs. Edmonds' "Blossom Algorithm" in Section 8.4 is an efficient algorithm for finding a maximum matching (Edmonds, 1965a). On the other hand, finding a maximum stable set is an NP-complete problem (Karp, 1972).

8.3 Min–Max Theorems

The first theorem in this section is motivated by a quaint Victorian puzzle called the "Marriage Problem."

> Given a set of men and a set of women, where each woman considers some of the men acceptable, under what conditions can each woman get married to an acceptable man?

A family of finite non-empty sets $S_1, S_2, \ldots, S_k$ has a *system of distinct representatives* (also called a *transversal*) if there exists a set $\{x_1, x_2, \ldots, x_k\}$ of distinct elements such that $x_1 \in S_1, x_2 \in S_2, \ldots, x_k \in S_k$.

For example, let $S_1 = \{1, 2, 3\}$, $S_2 = \{2, 4, 5\}$, and $S_3 = \{3, 5\}$. The situation can be modeled by the bipartite graph shown in Figure 8.4. Observe that $\{1, 4, 5\}$ is a transversal. It is not the only transversal; $\{1, 2, 3\}$ and $\{1, 3, 4\}$, for instance are also transversals.

Philip Hall determined the conditions under which a family of sets is guaranteed to have a transversal (Hall, 1935).

Figure 8.4 A system of distinct representatives.

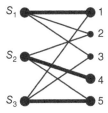

Theorem 8.3.1. (Marriage Theorem) *A family of finite non-empty sets* $S_1, S_2, \ldots, S_k$ *has a transversal if and only if the union of any j of these sets contains at least j elements, for every* $j \in \{1, \ldots, k\}$.

Construct a bipartite graph G with vertex classes X and Y, where $X = \{S_1, S_2, \ldots, S_k\}$ and $Y = S_1 \cup S_2 \cup \cdots \cup S_k$ as shown in Figure 8.4. For every $i \in \{1, \ldots, k\}$ draw an edge from vertex S_i in X to the elements of set S_i in Y. The Marriage Theorem is equivalent to finding a matching in G that covers all the vertices in X. It is this result and its proof that we will present.

Theorem 8.3.2. *Let G be a bipartite graph with vertex classes X and Y. Then G has a matching that covers X if and only if* $|N(A)| \geq |A|$, *for every* $A \subseteq X$.

Proof. Suppose G has a matching that covers X. Let $A \subseteq X$. Each edge of the matching takes an element of A to a different element in Y. So the size of the set of neighbors of vertices in A must be at least $|A|$. In other words $|N(A)| \geq |A|$.

Conversely, suppose for every $A \subseteq X$, $|N(A)| \geq |A|$. The proof is by induction on $|X| \geq 1$. The result is trivially true for $|X| = 1$ because $K_{1,n-1}$ has the required matching. Assume the result is true for all values less than $|X|$. Let G be a bipartite graph with vertex classes X and Y such that $|X| \geq 2$ and $|Y| \geq 2$.

First suppose that for every $A \subseteq X$, $|N(A)| > |A|$. Pick any vertex $x \in X$ and one of its neighbors $y \in Y$. Let H be the bipartite graph with vertex classes $X - \{x\}$ and $Y - \{y\}$, as shown in the following diagram. Observe that y may be adjacent to several vertices in X besides x, and in H the neighborhoods of all such vertices are reduced by 1. Due to the strict inequality $|N(A)| > |A|$, removal of vertices x and y leaves the bipartite graph H with the property that for every $A \subseteq X - \{x\}$, $|N(A)| \geq |A|$. By the induction hypothesis, H has a matching M_H covering all the vertices in $X - \{x\}$. Observe that M_H together with the edge connecting x and y is a matching in G covering all the vertices in X.

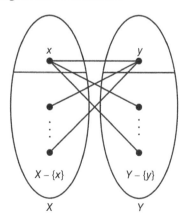

Next, suppose that for some $A \subset X$, $|N(A)| = |A|$. Without loss of generality let A be the largest such set. Let J be the subgraph of G with vertex classes $X - A$ and $Y - N(A)$. Since $|N(A)| = |A|$, X and Y lost the same number of vertices. By the maximality of A, for every $B \subseteq X - A$, $|N_J(B)| > |B|$. Applying the same argument as in the previous paragraph, we may conclude that J has a matching M_J that covers vertices in $X - A$. Let K be the subgraph of G with vertex classes A and $N(A)$. By the induction hypothesis K has a matching M_K that covers A. The matching $M_J \cup M_K$ is the required matching in G that covers all the vertices in X. □

Note that in the statement of Theorem 8.3.2 the phrase "a matching that covers X" can be replaced with $\beta(G) = |X|$ since G is a bipartite graph. We could write Theorem 8.3.2 as follows: For a bipartite graph G with vertex classes X and Y, $\beta(G) = |X|$ if and only if $|N(A)| \geq |A|$, for every $A \subseteq X$.

The next result is König's theorem that in a bipartite graph the size of the maximum matching is equal to the size of a minimum vertex cover (König, 1931). The following proof uses Theorem 8.3.2 and a simple set theoretic argument. However, König's theorem is equivalent to Theorem 8.3.2 (the proof is left as an exercise).

Theorem 8.3.3. (König's Matching Theorem) *Let G be a bipartite graph. Then $\beta(G) = \alpha'(G)$.*

Proof. Proposition 8.2.1 (i) implies that $\beta(G) \leq \alpha'(G)$ for any graph. We will prove that for a bipartite graph $\beta(G) \geq \alpha'(G)$; thereby proving that $\beta(G) = \alpha'(G)$. Let X and Y be the vertex classes of G and let C be a minimum vertex cover. Then $\alpha'(G) = |C|$. Let H be the induced subgraph formed with vertex classes $X \cap C$ and $Y - C$ and let K be the induced subgraph formed with vertex classes $X - C$ and $Y \cap C$. Then H and K are bipartite subgraphs and by their constructions, H and K have no overlapping edges. We must prove that H has a matching that covers $X \cap C$. Suppose, if possible, this is not true. Theorem 8.3.2 implies that there is a subset $A \subseteq X \cap C$ such that $|N_H(A)| < |A|$, as shown in the following diagram. Let

$$C' = (X \cap C - A) \cup N_H(A) \cup (Y \cap C).$$

Then C' is a vertex cover as shown by the highlighted region in the following diagram and

$$|C'| = |X \cap C| - |A| + |N_H(A)| + |Y \cap C| = |C| + |N_H(A)| - |A| < |C|.$$

This contradicts the minimality of C. Therefore H has a matching M_1 that covers $X \cap C$. Similarly K has a matching M_2 that covers $Y \cap C$. Then $M_1 \cup M_2$ is a matching in G. Therefore

$$\beta(G) \geq |M_1 \cup M_2| = |C| = \alpha'(G).$$

□

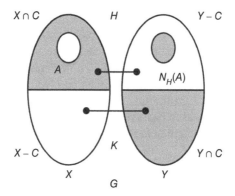

Next observe that for a bipartite graph, a large portion of the adjacency matrix has zeros. For example, the bipartite graph $K_{3,3}$ shown in Figure 1.8 with vertices in the first class labeled v_1, v_2, v_3 and those in the second class labeled v_4, v_5, v_6 has adjacency matrix

$$
A = \begin{array}{c} \\ v_1 \\ v_2 \\ v_3 \\ v_4 \\ v_5 \\ v_6 \end{array}
\begin{array}{c} v_1\ v_2\ v_3\ v_4\ v_5\ v_6 \\ \left(\begin{array}{cccccc} 0 & 0 & 0 & 1 & 1 & 1 \\ 0 & 0 & 0 & 1 & 1 & 1 \\ 0 & 0 & 0 & 1 & 1 & 1 \\ 1 & 1 & 1 & 0 & 0 & 0 \\ 1 & 1 & 1 & 0 & 0 & 0 \\ 1 & 1 & 1 & 0 & 0 & 0 \end{array}\right) \end{array} .
$$

The entire matrix is unnecessary because a small part of the matrix contains all the necessary information. Thus we define the *reduced adjacency matrix* of a bipartite graph $K_{r,s}$ with vertex classes $X = \{x_1, \ldots, x_r\}$ and $Y = \{y_1, \ldots, y_s\}$ as the $r \times s$ matrix A where $a_{ij} = 1$ if there is an edge between x_i and y_j with $x_i \in X$ and $y_j \in Y$. For example the reduced adjacency matrix of $K_{3,3}$ is given by

$$
A = \begin{array}{c} \\ v_1 \\ v_2 \\ v_3 \end{array}
\begin{array}{c} v_4\ v_5\ v_6 \\ \left(\begin{array}{ccc} 1 & 1 & 1 \\ 1 & 1 & 1 \\ 1 & 1 & 1 \end{array}\right) \end{array} .
$$

This reduced adjacency matrix A has some interesting properties:

(1) The ith row records the neighbors of x_i and the jth column records the neighbors of y_j. Therefore $deg(x_i) = \sum_j a_{ij}$ and $deg(y_j) = \sum_i a_{ij}$.

(2) The number of edges in G corresponds to the total number of ones in A.

(3) Labeling the vertices of X in a different order permutes the rows of A. Labeling the vertices of Y in a different order permutes the columns of A.

(4) If M is a matching in G, then M corresponds to a set of ones in A with no two ones in the same line (row or column). Thus $\beta(G)$ is the maximum number of ones in A with no two ones on a line.

(5) If C is a vertex cover in G, then C is a set of lines (rows and columns) of A needed to cover all the ones in A. Thus $\alpha'(G)$ is the minimum number of lines of A needed to cover all the ones in A.

Observe that any matrix consisting of zeros and ones corresponds to the reduced adjacency matrix of a bipartite graph. Therefore Theorem 8.3.3 can be stated entirely in terms of matrices. This version was independently obtained by Egerváry (1931).

Theorem 8.3.4. (Egerváry's Theorem)   *Let A be an $r \times s$ matrix consisting of zeros and ones. Then the maximum number of ones with no two ones in the same line (row or column) equals the minimum number of lines needed to cover all the ones of A.*

8.4   Maximum Matching Algorithm

The min–max theorems, elegant as they are, do not give methods for actually finding maximum matchings. The problem of finding a maximum matching (called the "Cardinality Matching Problem") came up in World War II. The Battle of Britain fought between July and October of 1940 pitted the Royal Air Force (RAF) against the German Air Force (Luftwaffe). The RAF assembled nearly 3000 pilots from across the British Empire and assigned two pilots who spoke the same language and had complemetary skills to a plane. The goal was to fly the largest number of planes per mission. The pilots available for a mission may be viewed as the vertices of a graph and two pilots were linked if they were suited for each other. A maximum matching in this graph corresponds to the most number of flights in the mission. (Roberts, 2005). Berge extended König and Hall's ideas by developing the augmenting path technique using which he gave an if and only if characterization for maximum matchings in any graph, not necessarily bipartite (Berge, 1957).

Let G be a graph and M be a matching. An *M-alternating path*[3] is a path whose edges alternate between M and $E(G) - M$. A vertex is *saturated* with respect to M if it is incident with an edge in M, otherwise it is called *unsaturated* with respect to M. An alternating path between two unsaturated vertices is called an *M-augmenting path*. By flipping or toggling (the preferred term) the matching, we can get a larger

---

3 Alternating paths originated with Petersen who used them to prove that a 3-connected cubic graph has a perfect matching (Petersen, 1891).

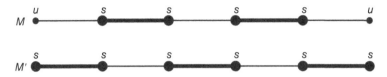

Figure 8.5 *M*-augmenting path.

matching M' as shown in Figure 8.5. More precisely, if we call the path P, thinking of P as a set of edges, then M' is the symmetric difference of P and M. In other words

$$M' = P \triangle M = (P - M) \cup (M - P).$$

For example, consider the graph in Figure 8.3 redrawn for convenience. Observe that $M = \{e_3, e_5\}$ is a matching that is not maximum. An M-augmenting path is $\{e_{11}, e_5, e_7, e_3, e_9\}$ as shown in the first diagram. The first and last vertices are not incident to any edge of M. This M-augmenting path gives a new matching $M' = \{e_{11}, e_7, e_9\}$, where $|M'| = |M| + 1$, as shown in the second diagram. Thus the presence of an M-augmenting path leads to a larger matching. On the other hand the matching $N = \{e_1, e_3, e_5\}$ is a maximum matching. It is straightforward to check that there is no N-augmenting path.

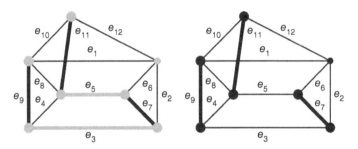

Theorem 8.4.1. (Berge's Maximum Matching Theorem) *Let G be a graph and M be a matching. Then M is a maximum matching if and only if G has no M-augmenting path.*

Proof. Suppose M is a maximum matching. Then G can have no M-augmenting path, since an M-augmenting path P gives rise to a matching $M' = M \triangle P$, where $|M'| = |M| + 1$.

Suppose G has no M-augmenting path and suppose, if possible, M is not a maximum matching. Then there is another matching M' such that $|M'| > |M|$. Construct a subgraph H such that $V(H) = V(G)$ and $E(H) = M \triangle M'$. Observe that the maximum degree of H is 2 since every vertex is incident to at most an edge in M

or an edge in M'. Thus H consists of isolated vertices, paths, and cycles. Further observe that the edges in a path will alternate between M and M'. Since $|M'| > |M|$, there must be a path that has more edges from M' than from M. This path will have two unsaturated vertices with respect to M, and is therefore an M-augmenting path. This is a contradiction to the hypothesis. □

Theorem 8.4.1 suggests a preliminary algorithm for finding maximum matchings. Given a graph G and a matching M, the algorithm searches for an M-augmenting path starting with an unsaturated vertex not incident with M. If an M-augmenting path is found, then using the procedure described in Theorem 8.4.1, it builds a larger matching and repeats the search for an M-augmenting path. If no M-augmenting path is found, then it stops and concludes that M is a maximum matching.

If the graph is bipartite then finding an augmenting path is straightforward, and this method works very well. The next algorithm taken from Schrijver (2003) illustrates how to find a maximum matching in a bipartite graph using the augmenting path technique in Theorem 8.4.1.

Algorithm 8.4.2. Bipartite Maximum Matching Algorithm

Input: A bipartite graph G with vertex classes X and Y and a matching M.
Output: A matching M' such that $|M'| > |M|$, if it exists.

(1) Turn G into a digraph as follows: For every arc $e = (x, y)$, where $x \in X$ and $y \in Y$, if $e \in M$, then orient e from y to x, otherwise orient e from x to y, as shown in the following diagram.

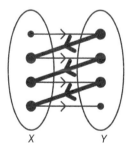

(2) Let X' and Y' be the unsaturated vertices in X and Y. An M-augmenting path (if it exists) is just a directed path from a vertex in X' to a vertex in Y'.

If G has n vertices and m edges, finding a directed path is $O(m)$, so finding an M-augmenting path is $O(m)$. Since this must be done for each vertex before concluding no M-augmenting path exists, the complexity of Algorithm 8.4.2 is $O(nm)$.

This algorithm might have been enough in an earlier era as noted in Witzgall (2001)

> The classical graph theorist would look at this elegant characterization of maximum matchings and ask: what more needs to be said? That outlook had been changing during and after World War II. The extensive planning needs, military and civilian, encountered during the war and post-war years now required finding actual solutions to many graph-theoretical and combinatorial problems, but with a new slant: Instead of asking questions about existence or numbers of solutions, there was now a call for "optimal" solutions, crucial in such areas as logistics, traffic and transportation planning, scheduling of jobs, machines, or airline crews, facility location, microchip design, just to name a few.

Searching for an M-augmenting path if the graph is not bipartite (or if it is not known that it is bipartite) is more complicated. Jack Edmonds' paper "Paths, trees, flowers" gives the early history of this topic and presents the beautiful Blossom Algorithm for finding maximum matchings in an arbitrary graph that is not necessarily bipartite (Edmonds, 1965a). Regarding Berge's technique, Edmonds wrote that Berge proposed tracing an alternating path from an unsaturated vertex until "it must stop." If the path is not augmenting he proposed "to back up a little and try again, thereby exhausting possibilities." However, as Edmonds noted, Berge's idea was an important improvement over "the completely naive algorithm." At that time (late nineteen fifties and early sixties) the notion of "efficient" had not yet crystallized. Edmonds was one of the first mathematicians to think about efficiency in algorithms. Pulleyblank (2012) describes how Edmonds thought of this idea as follows:

> The night before his scheduled talk, Edmonds had an inspiration with profound consequences. A graph is nonbipartite if and only if it has an odd cycle. It seemed that it was the presence of these odd cycles that confounded the search for augmenting paths. But if an odd cycle was found in the course of searching for an augmenting path in a nonbipartite graph, the cycle could be shrunk to form a pseudonode. Thereby the problem caused by that odd cycle could be eliminated, at least temporarily. This simple and elegant idea was the key to developing an efficient algorithm for determining whether a nonbipartite graph had a perfect matching. Equally important, it gave Edmonds a concrete specific example of a problem that could illustrate the richness and the power of the general foundations of complexity that he was developing. This became the focal point of his talk the next day which launched some of the most significant research into algorithms and complexity over the next two decades.

Figure 8.6 Matchings and blossoms.

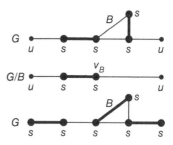

A *blossom* is the set of edges of an odd cycle with $2k + 1$ vertices, where $k \geq 4$, of which exactly k belong to a matching. Let G be a graph and M be a matching. A basic idea in Edmonds' argument is that if a blossom B is found in the process of looking for an M-augmenting path, then it is simply contracted down to one vertex v_B as shown in Figure 8.6. The matching $M - B$ in G/B is obtained by removing from M the matched edges of B. The search for an unsaturated vertex along an M-alternating path continues. If one is found, then the blossom is "uncontracted" and alternate edges of B are inserted into the M-augmenting path in G/B to obtain an M-augmenting path in G.

In addition to the idea of shrinking a blossom, Edmonds also needed a clever way of navigating through the graph. Any tree can be drawn as a *rooted tree* by picking one vertex and making it a root and drawing all the other edges hanging from it. Let G be a connected graph and M be a matching. An *M-alternating tree* T is a rooted tree, where the root is an unsaturated vertex and the edges along every path are alternately in M and not in M. Label the vertices of T even (e) or odd (o) depending on their distance from the root, which is always considered to be an even vertex. As a result edges labeled in order from the root as oe are in the matching M and edges labeled eo are not in M. Examples of M-alternating trees are shown in Figure 8.7.

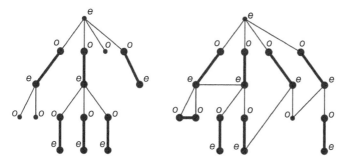

Figure 8.7 *M*-alternating trees.

Edges labeled e may have degree greater than two; however, edges labeled o can have degree at most 2. Otherwise suppose there is a vertex labeled o that has degree at least 3. Then there will be an edge labeled oe that is not in the matching M. To see this, imagine that the left most path in the first diagram in Figure 8.7 with vertices labeled in order as $eoeo$ has two edges attached to the last vertex labeled o. Then there will be two edges with end vertices labeled oe, but only one of them can be in the matching. Thus there will be an oe edge not in the matching, which is impossible.

Suppose an edge in G is incident to two vertices in an M-alternating tree T as shown in the right diagram of Figure 8.7. Then the edge together with the paths along T back to a common vertex forms a cycle. Moreover, this common vertex must be labeled e because vertices labeled o have degree at most 2 in T. There are two ways in which such a cycle forms a blossom:

(1) If an edge joins two vertices labeled e, then that edge forms an odd cycle with edges in the tree, and because the edges in the tree are M-alternating, the cycle is a blossom. Note that the edge joining the two even vertices cannot itself be part of the matching, as shown in Figure 8.7.
(2) If an edge in M joins two vertices labeled o, then the cycle formed has an odd number of M-alternating edges, and is therefore a blossom.

If an edge not in M joins two vertices labeled o, then it is still an odd cycle but does not have enough edges in M to be a blossom. An edge between two vertices that are not both labeled the same forms an even cycle and is not a blossom.

If a blossom occurs in a tree T, the path from the common vertex in the blossom to the root of T is called the *stem* of the blossom. The blossom together with its stem is called a *flower* as shown in Figure 8.8. The edge of the stem that is incident to the common vertex in the blossom is in M since it is labeled oe. Note again that a blossom contains an edge between two vertices labeled e or between two vertices labeled o (See figure 8.7). We will use this fact in the algorithm.

Suppose two odd vertices have an edge in the matching between them, as in the first graph in the following diagram. This edge, together with the paths from the two vertices back to a common vertex in the tree, forms a blossom. When a blossom B like this is encountered, we may contract it to a single vertex v_B and proceed with adding vertices and edges to the tree $T\backslash B$ in $G\backslash B$. If we reach an unsaturated vertex, which will be odd because all even vertices are saturated, then Theorem 8.4.1 implies that the path from the root of the tree to the unsaturated

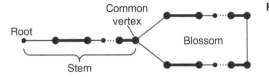

Figure 8.8 Flower.

vertex is an M-augmenting path, as shown in the second graph in the diagram. Let P be an M-augmenting path in G/B for the matching $M - B$. Let $M' = P\triangle(M - B)$ be the larger matching in G/B. In the original graph G, a path through B may be chosen to maintain the alternating pattern of edges in M and not in M and these edges may be added to M' to obtain a larger matching in G as shown in the third graph of the following diagram.

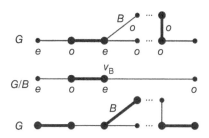

Next, suppose two even vertices have an edge between them as shown in the first graph in the following diagram. In this case the edge cannot be part of the matching, because even vertices are already saturated. This type of edge also forms a blossom B when taken together with the paths from its vertices back to a common vertex in the tree. The edges in this blossom are also alternately in M and outside M, and we can contract the blossom to a single vertex v_B and resume building the tree $T\backslash B$ in $G\backslash B$ as shown in the second graph. As with an edge between odd vertices, if there is an M-augmenting path in $G\backslash$, it can be extended to an M-augmenting path P in G by including a path through B that maintains the alternating pattern, and we replace M with $M\triangle P$ to obtain a larger matching, as shown in the third graph of the following diagram.

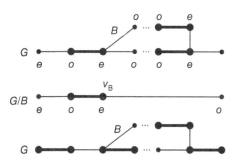

A *maximal M-alternating tree* is an M-alternating tree in which every leaf is labeled e and not adjacent to any M-saturated vertex outside the tree. Observe that only the root vertex in an M-alternating tree is unsaturated.

Algorithm 8.4.3. Edmond's Blossom Algorithm

Input: A graph G with a matching M.
Output: A maximum matching for G, a set of M-alternating trees, the set of even vertices of the trees and the set of odd vertices of the trees.

(1) Let $\mathcal{T}$ be an empty set of sets. Let E and O be empty sets of vertices.
(2) If G has no unsaturated vertices for M, then M is a maximum matching. Output $M, \mathcal{T}, E$, and O.
(3) Let T be an empty set of edges. Select an unsaturated vertex $u \notin E$. Let E_T and O_T be empty sets of vertices. Add u to E_T.
(4) **For each** $v \in E_T$:

> **For each** $v' \in N(v) - (E \cup E_t \cup O \cup O_T)$:
> Add v' to O_T.
> Add vv' to T.

(5) If any vertex $v \in O_T$ is unsaturated, then there is an M-augmenting path P in T from u to v. For any blossom B that has been contracted to a vertex v_B that lies on P, replace v_B with B and modify P to pass through B as an M-augmenting path. Replace M with $M \triangle P$. This will not affect any tree already in $\mathcal{T}$. If any unsaturated vertices remain, go back to step (3); otherwise M is a maximum matching, so output $M, \mathcal{T}, E$, and O.
(6) If there is any edge in M between two vertices in O_T, then the edge is part of a blossom B. Contract B to a new vertex v_B and continue the algorithm with $G/B, T/B, E_T - B$ and $O_T - B$. This will not affect any tree already in $\mathcal{T}$.
(7) Now every vertex $v \in O_T$ is incident to a distinct edge $vv' \in M$, where $v' \notin E \cup O$. (If $v' \in E \cup O$, then $v \in E \cup O$.) Add each of these edges to T and add each vertex v' to E_T.
(8) If there is any edge between two vertices in E_T, then the edge is part of a blossom B'. Contract B' to a new vertex $v_{B'}$ and resume the algorithm with $G/B', T/B', E_T - B'$ and $O_T - B'$. This will not affect any tree already in $\mathcal{T}$.
(9) Repeat steps (4) – (8) until either an M-augmenting path is found and M is replaced, or no more edges can be added to T.
(10) If steps (4) – (9) produce a tree T, add T to $\mathcal{T}$, add the vertices in E_T to E, and add the vertices in O_T to O. If any unsaturated vertex u remains, go to step (3).
(11) Output $M, \mathcal{T}, E$, and O.

The complexity of the algorithm is $O(n^2 m)$, where n is the number of vertices in the input graph and m is the number of edges. The number of unsaturated vertices requiring trees to be constructed is $O(n)$, and building each tree is $O(nm)$, due to iteration through vertices and matching updates when M-augmenting paths are found.

A tree returned by Algorithm 8.4.3 must be a maximal M-alternating tree. Furthermore, any tree node in O either is not saturated, in which case an M-augmenting path would have been returned instead of the tree, or is saturated, in which case it is incident to an edge in M which would be added to the tree, so that the node is not a leaf node. Therefore all leaf nodes of the tree are even. Finally, if a leaf node, which is even, is adjacent to an M-saturated vertex outside the tree, then the path from the root of the tree to that vertex would be an M-augmenting path, that would have been returned by the algorithm instead of the tree.

Theorem 8.4.4. *Let G be a graph with a matching M and a set of unsaturated vertices U. Algorithm 8.4.3 returns a maximum matching.*

Proof. Construct M, $\mathcal{T}$, E, and O as described in Algorithm 8.4.3. Every unsaturated vertex v is associated with a maximal M-alternating tree T_v. The root v is labeled e and its neighbors are all in T_v, and are all labeled o. If one of the neighbors is unsaturated, then it would have been added to the matching to obtain a larger matching by Algorithm 8.4.3 and the process would have begun again. So all the neighbors of v are saturated. Moreover, there is no M-alternating path between two vertices in different trees both labeled e. To see this, note that the M-alternating trees are maximal and when the M-alternating tree is constructed, any neighbor of a vertex labeled e would be added to the first constructed tree.

First consider the case when G is bipartite. Theorem 2.2.4 implies that every cycle in G has even length. In particular G has no blossom. Therefore there is no edge of the form ee or oo in the same tree. Suppose, if possible, G has an M-augmenting path P between two unsaturated vertices v and w, where v is the root vertex of the M-alternating tree T_v and w is the root vertex of the M-alternating tree T_w. Further let x be the last vertex in T_v along the path P and y is the first vertex in T_w along P, as shown in the following diagram.

Since x and y cannot both be labeled e, one of x or y must be labeled o with respect to its tree, say x is labeled o with respect to tree T_v. Then in the path from v to x

in T_v, the last edge $x'x$ must be an edge of the form eo and such edges are not in the matching. However, x is a saturated vertex, so there must be a vertex x'' in T_v such that xx'' is an edge in the matching of the form oe. Let x''' be the first vertex in P outside of T_v. Then clearly xx''' cannot also be in the matching. Therefore $x'x$ and xx''', as shown in the following diagram, are two adjacent unmatched edges in P; a contradiction since P is an M-augmenting path. Therefore M is a maximum matching.

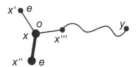

Next, consider the case when G is not bipartite. Suppose, if possible, G has an M-augmenting path P between unsaturated vertices v and w. As earlier, this path must pass through the M-alternating trees T_v and T_w constructed by the algorithm. Since G is not bipartite, one or more blossoms may have been encountered and contracted during construction of T_v and T_w. Any M-augmenting path in G not passing through a blossom would still be an M-augmenting path with the blossom contracted, which cannot exist by the same argument as in the bipartite case. Any M-augmenting path through a blossom must pass through its common vertex, by construction, and therefore would still be an M-augmenting path in G/B. Again, by the same argument as in the bipartite case, this cannot happen. Therefore G does not have any M-augmenting paths and so M is maximum. □

We will illustrate Algorithm 8.4.3 using the graph in Figure 8.9, with the indicated matching $M = \{bc, fk, hl, in, rs\}$.

- In the beginning $\mathcal{T}$, E, and O are empty. The unsaturated vertices in the graph are $a, d, e, g, j, m, o, p, q,$ and t. Begin by building an M-alternating tree starting

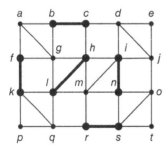

Figure 8.9 A graph with a matching.

with vertex m. At Step 3 in the algorithm, start with an empty tree T, an empty set of vertices O_T, and a set of vertices E_T containing only m.

- Since m has neighbors h, i, l, n, and r, in Step 4 add each of these vertices to O_T, and add edges hm, im, lm, mn, and mr to T.
- At Step 5, since none of the vertices in O_T are unsaturated, we proceed.
- At Step 6, since $lh \in M$ is an edge between two vertices in O_T, there is a blossom B consisting of edges ml, lh, and hm. Contract this blossom to a new vertex v_B, as shown in the following diagram. After B is contracted, $E_T = \{v_B\}$, $O_T = \{i, n, r\}$, and $T = \{iv_B, v_Bn, v_Br\}$.

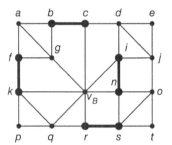

- Next, still at Step 6, since $ni \in M$ is another edge connecting two vertices in O_T, there is another blossom B' consisting of edges v_Bn, ni and iv_B. We contract this blossom to a new vertex $v_{B'}$, as shown in the following diagram. After B' is contracted, $E_T = \{v_{B'}\}$, $O_T = \{r\}$ and $T = \{v_{B'}r\}$. There are no more blossoms at this point to contract, so we proceed to Step 7.

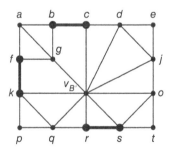

- At Step 7, since $rs \in M$ is incident with r, add s to E_T and add rs to T. Now $E_T = \{v_{B'}, s\}$, $O_T = \{r\}$, and $T = \{v_{B'}r, rs\}$.

- At Step 8, we observe that the two vertices in E_T, $v_{B'}$ and s, are adjacent, and so there is a blossom B'' consisting of edges $v_{B'}r$, rs, and $sv_{B'}$. We contract this blossom to a new vertex $v_{B''}$ as shown in the following diagram. After B'' is contracted, $E_T = \{v_{B''}\}$, and O_T and T are both empty.

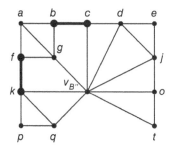

- Returning to Step 4, observe that $v_{B''}$, the only vertex in E_T, has neighbors c, d, g, j, k, o, q, and t. We add each of these vertices to O_T and add edges $cv_{B''}$, $dv_{B''}$, $gv_{B''}, jv_{B''}, ov_{B''}, qv_{B''}$, and $tv_{B''}$ to T.
- At Step 5, observe that since d is an unsaturated vertex, there is an M-alternating path $v_{B''}d$ in the contracted graph. Since this path touches a contracted blossom $v_{B''}$, remove the contraction. In the resulting graph $v_{B'} - d$ is an M-augmenting path. Remove the contraction at $v_{B'}$ as well, and note that, as indicated in the proof of the algorithm, we may continue the path around the blossom to find that $d - i - n - v_B$ is an M-augmenting path. Finally, remove the contraction at v_B, and we can see that $d - i - n - m$ is an M-augmenting path in the original graph. We replace ni in the matching with mn and di. The resulting graph with a larger matching is shown in Figure 8.10.
- Select a different unsaturated vertex and start the process over until every unsaturated vertex is the root of an M-alternating tree.

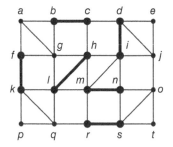

Figure 8.10 A graph with a larger matching.

Edmonds work on efficient algorithms has been referred to as a "glimpse of heaven."(Edmonds, 1991) We will end the book with this glimpse.

Exercises

8.1 Find $\alpha(G)$, $\alpha'(G)$, $\beta(G)$, and $\beta'(G)$ for the graphs in Figures 1.33, 2.8, 3.3, and 3.4.

8.2 Apply Algorithm 8.1.5 to the strongly connected weighted digraph in Figure 5.6, using a as the source and k as the sink.

8.3 Let G be a bipartite graph with no isolated vertices.
(a) Prove Theorem 8.3.2 (Marriage Theorem) using Theorem 8.3.3 (König's Bipartite Matching Theorem).
(b) Prove Theorem 8.3.2 using Theorem 8.4.1 (Berge's Augmenting Path Theorem)

8.4 Let G be a bipartite graph with vertex classes X and Y. Prove that:

$$\beta(G) = |X| - \max_{S \subset X} \{|A| - |N(A)|\}.$$

8.5 Let G be a graph with n vertices, where n is even. Prove that
(a) If G is a regular bipartite graph, then G has a perfect matching.
(b) If the degree of every vertex is at least $\frac{n}{2}$, then G has a perfect matching.
(c) If G is a bridgeless cubic graph, then G has a perfect matching (Petersen, 1891).
(d) If any $k \leq \frac{3n}{4}$ vertices are adjacent to at least $\frac{4k}{3}$ vertices, then G has a perfect matching (Anderson, 1971).

8.6 Let G be a bipartite graph with no isolated vertices. Prove that G has a perfect matching if and only if the number of isolated vertices in $G - A$ is at most $|A|$, for every $A \subseteq V(G)$.

8.7 A component of a graph is called *odd* if it has an odd number of vertices. Denote the number of odd components of $G - A$ as $odd(G - A)$. Prove that a graph has a perfect matching if and only if $odd(G - A) \leq |A|$ for every $A \subset V(G)$ (Tutte, 1947). This is called Tutte's Perfect Matching Theorem. It characterizes perfect matchings in any graph, not necessarily bipartite. Tutte's original proof is matrix theoretic. See Anderson's proof that uses Theorem 8.3.2 and an induction argument (Anderson, 1971), See also Lovász, (1975).

8.8 Let G be a graph on n vertices none of which are isolated. Prove that

$$\beta(G) = \frac{1}{2} \left(n - \max_{A \subseteq V(G)} \{ odd(G - A) - |A| \} \right).$$

This result is known as "Tutte-Berge Formula for Maximum Matchings" (Berge, 1958).

Topics for Deeper Study

8.9 The Marriage Problem is more generally called the *Assignment Problem* on bipartite graphs and stated in the following manner: Given a set of r workers and s jobs, where each worker is suited for some of the jobs, is there an assignment of workers to jobs so that each worker gets a job she is suited for? Review applications of the Assignment Problem in operations research.

8.10 The optimization problem for a weighted graph corresponding to the Assignment Problem for a graph is called the *Optimal Assignment Problem*. It is stated as follows: Given a set of r workers and s jobs, where worker i's potential for job j is x_{ij}, is there an assignment of workers to jobs so as to maximize $\sum x_{ij}$? A *maximum weight matching M* in a weighted bipartite graph G is a matching with the property that the sum of the weights on the edges in M is largest among all possible maximum matchings. Harold Kuhn gave an algorithm known as the Hungarian Algorithm (Kuhn, 1955). Review this algorithm.

8.11 Let G be a complete bipartite graph with vertex classes W and M. Suppose W consists of women and M consists of men and each woman ranks all the men and each man ranks all the women in order of preference. We say a maximum matching M is *stable* if there is no pairing wm and $w'm'$ in which w ranks m' higher than m and m ranks w' higher than w. The goal is to find a stable matching in G. David Gale and Lloyd Shapley gave an efficient algorithm for this problem, known as the *Stable Marriage Problem* (Gale and Shapley, 1962). Review this algorithm and its applications.

8.12 Edmonds also gave a polynomial-time algorithm for the maximum weight matching problem on a weighted graph that is not necessarily bipartite. Review this algorithm. See for example Schrijver (2003).

8.13 Recall the Postman Problem from Section 6.1 that asked for the minimum number of edges to Eulerize a graph. The maximum weight matching problem can be turned into a minimum weight matching problem (Edmonds, 1965b). Let G be a connected graph and let O be the set of odd vertices in G. The size of O is even. Calculate the distance between every pair of vertices in O and call this a weight function d. Construct the complete weighted graph with vertex set O and weight function d. Find a minimum weight perfect matching M of this complete weighted graph. It will have $\frac{|O|}{2}$ edges. For each edge xy in M, find a corresponding $x - y$ path in G. The edges in those paths are the ones that must be traversed twice to Eulerize the graph. Review this method.

8.14 Let G be a graph with n vertices none of which are isolated. We say $D \subseteq V(G)$ is a *dominating set* if every vertex in $V(G) - D$ is adjacent to at least one vertex in D. The number of vertices in the smallest dominating set is denoted by $dom(G)$ and called the *dominating number* of G (Haynes et al., 1998).

(a) Find the dominating number of the graphs in Figures 1.33, 2.8, 3.3, and 3.4.

(b) Let G be a connected graph with $\delta \geq 3$. Prove that $dom(G) \leq \frac{3\delta}{2}$ (Reed, 1996).

(c) A *stable dominating set* is a dominating set that is also a stable set. The number of vertices in the smallest stable dominating set is denoted by $i(G)$. Prove that if G has no induced subgraph isomorphic to $K_{1,3}$, then $i(G) = dom(G)$ (Allan and Laskar, 1978).

(d) A subset of vertices X is a *total dominating set* if every vertex in G is adjacent to a vertex in X. Prove that if $\delta \geq 3$, then $|X| \geq \frac{n}{2}$ (Archdeacon et al., 2004).

8.15 Let G be a graph with vertices $\{v_1, \dots, v_n\}$. Zero forcing is a coloring game, where initially each vertex of G is colored blue or green and the goal is to color all n vertices blue by applying the color change rule: a blue vertex b can change the color of a green vertex r to blue if r is the unique green neighbor of b. The minimum number of blue vertices needed to color all the vertices blue is called the *zero forcing number* $Z(G)$. Practically speaking, zero forcing can be used to determine where to place units that measure voltage and current in an electric grid so as to cover the entire grid. See Fallat et al. (2020).

(a) Find the zero forcing number of the graphs in Figures 1.33, 2.8, 3.3, and 3.4.

(b) Let $S(G)$ be the set of all real symmetric matrices $M = [a_{ij}]$, where a_{ij} is non-zero if $v_i v_j$ is an edge and 0 otherwise. The inverse eigenvalue problem for graphs asks what possible eigenvalues can this family of matrices have? The zero forcing number is an upper bound for the maximum multiplicity of an eigenvalue of any matrix in $S(G)$. Review this topic and find the zero forcing number of paths, cycles, wheels, and stars.

Appendix A

Linear Algebra

A vector $\bar{x}$ is a list of numbers. A matrix is a two-dimensional array of numbers with m rows and n columns. If $m = n$, we call the matrix a square matrix. If λ is a real number and A is an $m \times n$ matrix, then λA is obtained by multiplying each entry of A by λ. If A and B are $m \times n$, then the sum $A + B$ is an $m \times n$ matrix obtained by adding corresponding entries. If A and B are $m \times p$ and $p \times n$ matrices, respectively, then the product AB is an $m \times n$ matrix whose entries are defined as

$$c_{ij} = \sum_{k=1}^{p} a_{ik} b_{kj},$$

where a_{ij}, b_{ij}, and c_{ij} denote the entries in row i and column j of A, B, and AB, respectively.

The *transpose* of an $m \times n$ matrix A, denoted by A^T, is the matrix with rows and columns switched. An $n \times n$ matrix A is called *symmetric* if $A^T = A$.

The *inverse* of an $n \times n$ matrix A, if it exists, is the matrix A^{-1}, such that $AA^{-1} = A^{-1}A = I_n$, where I_n is the $n \times n$ matrix with ones on its diagonal and zeroes elsewhere, called the identity matrix. In this case A is called *invertible*.

The *first minor* of an $m \times n$ matrix A, denoted by M_{ij}, is the submatrix obtained by deleting the ith row and jth column. The *determinant* of an $n \times n$ matrix $A = [a_{ij}]$ is defined recursively as

$$det(A) = \sum_{j=1}^{n} (-1)^{j+1} a_{1j} det(M_{1j}).$$

Observe that $det(A) = det(A^T)$. An $n \times n$ matrix A is invertible if and only if $det(A) \neq 0$. Moreover, the matrix equation $A\bar{x} = \bar{0}$ has a non-zero solution if and only if A is not invertible.

An *eigenvector* of an $n \times n$ matrix A is a non-zero vector $\bar{x}$ such that $A\bar{x} = \lambda\bar{x}$, for some real number λ. We call λ the *eigenvalue* corresponding to eigenvector $\bar{x}$.

Graphs and Networks, First Edition. S. R. Kingan.
© 2022 John Wiley & Sons, Inc. Published 2022 by John Wiley & Sons, Inc.

Observe that $A\bar{x} = \lambda\bar{x}$ gives $(A - \lambda I_n)\bar{x} = \bar{0}$. The matrix equation

$$(A - \lambda I_n)\bar{x} = \bar{0}$$

has non-zero solutions if and only if $det(A - \lambda I_n) = 0$. This polynomial equation in λ is called the *characteristic polynomial* of A. To find the eigenvalues find the roots of the characteristic polynomial. Once the eigenvalues are found, the eigenvectors corresponding to each distinct eigenvalue λ are obtained by solving the matrix equation $(A - \lambda I_n)\bar{x} = 0$. The set of solutions to $(A - \lambda I_n)\bar{x} = 0$ is called the eigenspace corresponding to λ.

For example, the adjacency matrix A of the cycle on 3 vertices C_3 is

$$A = \begin{bmatrix} 0 & 1 & 1 \\ 1 & 0 & 1 \\ 1 & 1 & 0 \end{bmatrix}.$$

Therefore

$$A - \lambda I_3 = \begin{bmatrix} -\lambda & 1 & 1 \\ 1 & -\lambda & 1 \\ 1 & 1 & -\lambda \end{bmatrix}.$$

Observe that

$$det(A - \lambda I_3) = -\lambda(\lambda^2 - 1) - 1(-\lambda - 1) + 1(1 + \lambda) = -\lambda^3 + 3\lambda + 2.$$

Solving the polynomial equation $-\lambda^3 + 3\lambda + 2 = 0$ gives

$$\lambda = -1, -1, 2.$$

The eigenvector corresponding to $\lambda = 2$ is obtained by solving the matrix equation

$$\begin{bmatrix} -2 & 1 & 1 \\ 1 & -2 & 1 \\ 1 & 1 & -2 \end{bmatrix} \begin{bmatrix} x_1 \\ x_2 \\ x_3 \end{bmatrix} = \begin{bmatrix} 0 \\ 0 \\ 0 \end{bmatrix},$$

to get

$$\bar{x} = x_3 \begin{bmatrix} 1 \\ 1 \\ 1 \end{bmatrix},$$

where x_3 is a real number. The eigenvector corresponding to $\lambda = -1$ is obtained by solving the matrix equation

$$\begin{bmatrix} 1 & 1 & 1 \\ 1 & 1 & 1 \\ 1 & 1 & 1 \end{bmatrix} \begin{bmatrix} x_1 \\ x_2 \\ x_3 \end{bmatrix} = \begin{bmatrix} 0 \\ 0 \\ 0 \end{bmatrix},$$

to get

$$\bar{x} = \begin{bmatrix} -x_2 - x_3 \\ x_2 \\ x_3 \end{bmatrix} = x_2 \begin{bmatrix} -1 \\ 1 \\ 0 \end{bmatrix} + x_3 \begin{bmatrix} -1 \\ 0 \\ 1 \end{bmatrix},$$

where x_2 and x_3 are real numbers.

The next proposition lists some useful results on eigenvalues.

Proposition A.1.   *Let A be a square matrix with eigenvalue λ. Then*

(i) *$c\lambda$ is an eigenvalue for cA, where c is a non-zero constant;*
(ii) *λ^k is an eigenvalue of A^k;*
(iii) *λ is an eigenvalue of A^T; and*
(iv) *A is non-invertible if and only if 0 is an eigenvalue.*

Proof.   To prove (i) observe that

$$(cA)\bar{x} = c(A\bar{x}) = c(\lambda\bar{x}) = (c\lambda)\bar{x}.$$

Therefore $c\lambda$ is an eigenvalue of cA.

The proof of (ii) requires an induction argument. The result holds for $n = 1$ since $A\bar{x} = \lambda\bar{x}$. Assume the result holds for A^{k-1} and consider A^k. By the induction hypothesis

$$A^k\bar{x} = A^{k-1}(A\bar{x}) = A^{k-1}\lambda\bar{x} = \lambda A^{k-1}\bar{x} = \lambda\lambda^{k-1}\bar{x} = \lambda^k\bar{x}.$$

To prove (iii) observe that

$$det(A - xI_k) = det\left((A - xI_k)^T\right) = det\left(A^T - \bar{x}I_k\right).$$

Therefore A and A^T have the same eigenvalues.

To prove (iv) observe that 0 is an eigenvalue of A if and only if the matrix equation $A\bar{x} = 0\bar{x}$ holds for a non-zero vector $\bar{x}$, and this happens if and only if A is non-invertible.   □

Two matrices A and B are *similar* if there exists an invertible matrix C such that $B = CAC^{-1}$. It can be shown that similar matrices have the same characteristic polynomial, and therefore the same eigenvalues.

Proposition A.2.   *Similar matrices have the same eigenvalues.*

A square matrix that has non-zero entries only on its diagonal is called a *diagonal matrix*. A square matrix A is called *diagonalizable* if A is similar to a diagonal matrix (*i.e.* $A = CDC^{-1}$ for a diagonal matrix D and an invertible matrix C).

The *inner product* of two vectors $\bar{u} = [u_1, u_2, \ldots, u_n]$ and $\bar{v} = [v_1, v_2, \ldots, v_n]$ is defined as

$$\bar{u} \cdot \bar{v} = u_1 v_1 + \cdots + u_n v_n.$$

The length of $\bar{v}$ is defined as

$$\|\bar{v}\| = \sqrt{v_1^2 + \cdots + v_n^2}.$$

If $\|\bar{v}\| = 1$ we call the vector a unit vector. Two vectors $\bar{u}$ and $\bar{v}$ are *orthogonal* if $\bar{u} \cdot \bar{v} = 0$. If, in addition, they are unit vectors, they are called *orthonormal*.

A square matrix A in which the column vectors are pairwise orthogonal is called an *orthogonal matrix*. If, in addition, the columns are unit vectors, then it is called an *orthonormal matrix*. It this case it can be shown that $A^{-1} = A^T$. The next result appears in any Linear Algebra textbook. See, for example, example, Lay (2016).

Theorem A.3. (Diagonalization of Symmetric Matrices) *Let A be a square matrix. Then A is symmetric if and only if $A = CDC^T$, where D is a diagonal matrix whose entries are the eigenvalues of A and C is an orthonormal matrix.*

A symmetric matrix consisting of real numbers (called a real symmetric matrix) has nice properties, many of which follow directly from Theorem A.3. Two such properties are listed in the next proposition. The *trace* of an $n \times n$ matrix A, denoted by tr(A), is the sum of the entries of the diagonal.

Proposition A.4. *Let A be a real symmetric matrix. Then*

 (i) The eigenvalues of A are real numbers; and
(ii) tr(A) is the sum of the eigenvalues of A.

Proof. The proof of (i) is an immediate consequence of Theorem A.3. The proof of (ii) follows from Theorem A.3 and the fact that tr(A) = tr(PDP^{-1}) = tr(D).   □

Appendix B

Probability and Statistics

In statistics a population is the entire set of objects, individuals, or events under study and a sample is a subset of the population. A parameter is a numerical value that summarizes the population data and a statistic is a numerical value that summarizes the sample data. In probability theory the set of all possible outcomes of an experiment is called the sample space and a subset of the sample space is called an event. Population in statistics corresponds to sample space in probability theory and sample in statistics corresponds to event in probability theory. A *random variable*, typically denoted by X, is a real-valued function that measures the outcome of an experiment or observation. A random variable that takes a countable number of values is called a discrete random variable and one that takes uncountable number of values is called a continuous random variable. The probability distribution of a random variable is a listing of the probability of occurrence for each value of the random variable.

Probability functions are used to model the distributions of empirical random variables. The terminology follows Miller and Miller (2014). A *discrete probability function* is a function $f(x)$ defined over a countable set of real numbers such that for all $x, f(x) \geq 0$ and

$$\sum_x f(x) = 1.$$

A *continuous probability function* is a function $f(x)$ defined over the set of real numbers such that for all $x, f(x) \geq 0$ and

$$\int_{-\infty}^{\infty} f(x)\, dx = 1.$$

A discrete probability function $f(x)$ is used to model the probability distribution of a discrete random variable X. We write

$$P(X = x) = f(x).$$

Graphs and Networks, First Edition. S. R. Kingan.
© 2022 John Wiley & Sons, Inc. Published 2022 by John Wiley & Sons, Inc.

A continuous probability function $f(x)$ is used to model the probability distribution of a continuous random variable X. We write

$$P(a \leq X \leq b) = \int_a^b f(x)\, dx.$$

The *mean* (also called expectation) of a random variable X with probability function $f(x)$ is defined as

$$\mu = \sum x f(x),$$

if $f(x)$ is discrete and

$$\mu = \int_{-\infty}^{\infty} x f(x)\, dx,$$

if $f(x)$ is continuous. The *variance* is defined as

$$\sigma^2 = \sum (x - \mu)^2 f(x),$$

if $f(x)$ is discrete and

$$\sigma^2 = \int_{-\infty}^{\infty} (x - \mu)^2 f(x)\, dx,$$

if $f(x)$ is continuous. The *standard deviation* is the square root of the variance.

The binomial, normal, and Poisson probability functions are three popular and frequently used probability functions.

(1) The *binomial probability function* is a discrete function with parameters n and p, where $0 \leq p \leq 1$. It is defined as

$$B(x; n, p) = \binom{n}{x} p^x (1 - p)^{n-x}.$$

The mean of this function is $\mu = np$ and standard deviation is $\sigma = \sqrt{np(1 - p)}$. For example, consider the experiment of tossing two fair coins. The sample space is the set $\{HH, HT, TH, TT\}$, where H and T denote the outcomes of getting heads and tails, respectively. Let X be the number of heads. Then X is a discrete random variable that takes the values $0, 1, 2$. The probability distribution of X is modeled by the binomial probability with $n = 2$ and $p = 0.5$. In general, any process that consists of n trials, where each trial has two outcomes, success with probability p and failure with probability $1 - p$, is modeled by the binomial distribution.

(2) The *normal probability function* is a continuous function, with parameters μ and $\sigma > 0$. It is defined as

$$\mathcal{N}(x; \mu, \sigma) = \frac{1}{\sigma\sqrt{2\pi}} e^{-\frac{(x-\mu)^2}{2\sigma^2}}.$$

The mean is μ and standard deviation is σ.

(3) The *Poisson probability function* is a discrete probability function, with parameter $\lambda > 0$. It is defined as

$$P(k; \lambda) = \frac{\lambda^k e^{-\lambda}}{k!}.$$

The mean and standard deviation are both λ.

These three probability functions are fundamentally linked to each other. Consider a binomial function $\mathcal{B}(x; n, p)$. As the number of trials, n, tends to infinity, it may be approximated by the normal function $\mathcal{N}(np, np(1 - p))$. As n tends to infinity and the probability of success p tends to 0, the product np may be viewed as a constant $\lambda = np$. In this case, the binomial function may be approximated by the Poisson function $P(x; \lambda)$. The Poisson function is a good approximation for the binomial function when $n \geq 100$ and $np < 10$. The Poisson function also models situations when X measures the number of successes that occur in a specific time interval or region.

Let X and Y be two discrete random variables. A discrete bivariate real-valued function $f(x, y)$ such that $f(x, y) \geq 0$ and

$$\sum_x \sum_y f(x, y) = 1$$

is called a *joint probability function*. A bivariate probability function $f(x, y)$ is used to model the joint probability distribution of X and Y. We write

$$P(X = x, Y = y) = f(x, y).$$

Suppose X and Y are two random variables with means μ_X and μ_Y, respectively, standard deviations σ_X and σ_Y, respectively, and joint probability distribution p_{ij}. The covariance of X and Y is defined as

$$\sigma_{XY} = \sum_i (x_i - \mu_x)(y_i - \mu_Y)p_{ij}.$$

The (Pearson) *correlation coefficient* is defined as

$$r = \frac{\sigma_{XY}}{\sigma_X \sigma_Y}.$$

The correlation coefficient is a measure of the strength of association between X and Y. Note that $-1 \leq r \leq 1$. A value of r closer to 1 indicates a strong positive association and a value closer to -1 indicates a strong negative association. A value closer to 0 indicates little or no association.

Appendix C

Complexity of Algorithms

An algorithm is a sequence of computational operations that transforms input values into output values. The *worst-case complexity* of an algorithm, often referred to simply as the complexity, is the maximum number of steps it takes to transform the input data into the output data. For a graph with n vertices and m edges, it is usually a function of n or m or both. We characterize an algorithm's complexity function by its rate of growth relative to certain simply expressed functions, such as a polynomial function $f(x) = x^k$, where k is a fixed natural number; an exponential function $f(x) = b^x$, where $b > 0$ is a real number; a logarithm function $f(x) = \log_b x$; or a factorial function $f(x) = x!$, where x is a natural number. Eventually the exponential function grows faster than any polynomial function and its inverse the log function grows the slowest. The factorial function grows even faster than the exponential function.

The terminology in this appendix follows Cormen et al. (2001). Suppose $f(n)$ and $g(n)$ are functions from the set of natural numbers to the set of positive real numbers that express the worst-case complexity of two different algorithms in terms of input size n.

If for some real number $K > 0$, there exists a natural number n_0, such that for all $n \geq n_0$, $f(n) \leq Kg(n)$, then we say f is *asymptotically bounded above* by g. The set of all functions asymptotically bounded above by g is denoted as $O(g)$. If f is in the set $O(g)$, it is convention to say f **is** $O(g)$.

If for some real number $K > 0$, there exists a natural number n_0, such that for all $n \geq n_0$, $f(n) \geq Kg(n)$, then we say f is *asymptotically bounded below* by g. The set of all functions asymptotically bounded below by g is denoted as $\Omega(g)$. If f is asymptotically bounded below by g, we say f **is** $\Omega(g)$.

If f is asymptotically bounded above by g and also asymptotically bounded below by g, we say f and g are *asymptotically tightly bound*. The set of all functions that are asymptotically tightly bound with g is called $\Theta(g)$. If f and g are tightly bound, we say f **is** $\Theta(g)$. By definition, f is $\Theta(g)$ if and only if g is $\Theta(f)$.

Graphs and Networks, First Edition. S. R. Kingan.
© 2022 John Wiley & Sons, Inc. Published 2022 by John Wiley & Sons, Inc.

For example, let $f(n) = 3n^2 + 4n + 1$ and $g(n) = n^2$. Then f is asymptotically bounded above by g since $f(n) \leq 4g(n)$ for $n \geq 6$, and asymptotically bounded below by g since $f(n) \geq g(n)$ for $n \geq 1$. Hence f is $\Theta(g)$.

All polynomial functions of the same degree are asymptotically tightly bound to each other. We say that a polynomial function of degree k is $\Theta(n^k)$.

All logarithm functions are asymptotically tightly bound to each other. Let $f(n) = \log_a n$ and $g(n) = \log_b n$, where $a, b > 0$. Then f and g are asymptotically tightly bound. To see this, observe that $f(n) = \log_a n = \frac{\ln n}{\ln a}$ and $g(n) = \frac{\ln n}{\ln b}$, and therefore

$$f(n) = \frac{\ln b}{\ln a} g(n).$$

However, two exponential functions $f(n) = a^n$ and $g(n) = b^n$, where $a, b > 0$ are asymptotically tightly bound if and only if $a = b$.

We can express a stronger relationship between complexity functions by replacing the term "some" in the previous definitions with the term "every." If for every real number $K > 0$, there exists a natural number n_0, such that for all $n > n_0$, $f(n) \leq Kg(n)$, we say that f has lower order than g. The set of all functions with lower order than g is denoted $o(g)$. If f has lower order than g, we say f is $o(g)$. In this case f is of lower order than g if and only if

$$\lim_{n \to \infty} \frac{f(n)}{g(n)} = 0.$$

For example, let $f(n) = k^n$, where k is a natural number, and let $g(n) = n!$. Let

$$a_n = \frac{f(n)}{g(n)} = \frac{k^n}{n!}.$$

Then

$$\frac{a_{n+1}}{a_n} = \frac{k^{n+1}}{(n+1)!} \frac{n!}{k^n} = \frac{k}{n}.$$

When computing limits, k is constant and n is increasing, and when $n > 2k$, $\frac{k}{n} < \frac{1}{2}$. Thus when $n > 2k$, $0 < a_{n+1} < \frac{1}{2} a_n$, so $\lim_{n \to \infty} a_n = 0$. Therefore an exponential function has lower order than the factorial function.

Note that if f is $o(g)$, then f is $O(g)$. However, the converse is not true. For example, let $f(n) = n$ and $g(n) = 2n$. Then for all $n \geq 1$, $f(n) \leq 2g(n)$, so f is $O(g)$. However, $\lim_{n \to \infty} \frac{f(n)}{g(n)} = \frac{1}{2}$, so f is not $o(g)$.

Similarly, we say that f has higher order than g if

$$\lim_{n \to \infty} \frac{f(n)}{g(n)} = \infty.$$

The set of all functions with higher order than g is called $\omega(g)$, and if $f \in \omega(g)$ we say that f is $\omega(g)$. Note that if f is $\omega(g)$, then f is $\Theta(g)$. However, the converse is not true. For example, let $f(n) = n + (-1)^n$ and $g(n) = n$. Then f is $\Theta(g)$. However, $\lim_{n \to \infty} \frac{f(n)}{g(n)}$ is undefined, so f is not $\omega(g)$.

The notation $O(g)$ appears more frequently in computer science literature than any of the other notations since typically we are more interested in determining an upper bound on the worst-case behavior of an algorithm. It is also much easier in many cases to prove that the worst-case behavior of an algorithm is bounded above by a function than to prove that it has a tight asymptotic relationship. In the previous example where $g(n) = n^2$ and $h(n) = 3n^2 + 4n + 1$, it is not wrong to say h is $O(g)$, since n^2 has the biggest impact on the growth of h. Similarly, if $g(n) = n^2$ and $t(n) = 2^n$, it is not wrong to say that g is $O(t)$, but arguably pointless.

An algorithm whose complexity function is $O(n)$ is a *linear-time algorithm*. An algorithm whose complexity function is $O(n^k)$ for some natural number k is called a *polynomial-time algorithm*. An algorithm whose complexity function is $O(\log n)$ is called a *log-time algorithm*, and so on. An algorithm with a constant complexity function is called an $O(1)$ *algorithm*.

We define a *problem P* to be a binary relation on a set I called problem instances and a set S called problem solutions. A pair $(i, s) \in I \times S$ is in P if s is a solution to the problem for instance i.

For example, consider the problem of finding Hamiltonian cycles. The set I consists of graphs, the set S has two values, say, true and false, and

$$P = \{(G, s) \in I \times S \mid s \text{ is true if and only if } G \text{ has a Hamiltonian cycle}\}.$$

For the Traveling Salesman Problem the set I is the set of complete graphs with a cost function on $V \times V$ to the set of natural numbers, the set S is the set of natural numbers, and

$$P = \{(N, c) \in I \times \mathbb{Z} \mid c \text{ is the } minimum \text{ total cost of a Hamiltonian cycle}\}.$$

Note that the Hamiltonian cycle that gives the minimum cost would be computed as part of any algorithm to solve the problem.

A *decision problem* is a problem with a yes or no answer that depends on the input. Determining whether or not a cycle is Hamiltonian cycle is a decision problem. An *optimization problem* is a problem in which each solution has a number associated with it and the goal is to find the solution with the "best" value. The Traveling Salesman Problem is an optimization problem.

We can usually convert an optimization problem into a decision problem by imposing an upper bound on the value to be optimized. For example, to convert the Traveling Salesman Problem into a decision problem, we ask whether the graph has a Hamiltonian cycle of cost at most k. Thus if the optimization problem is easy, then its related decision problem is easy. Equivalently, if the decision problem is difficult, then so is the optimization problem.

A *deterministic sequential machine* is a machine that performs only one operation in an instant and performs operations sequentially. A *non-deterministic machine* can perform several operations in an instant.[1] The set of decision problems that can be solved in polynomial time on a deterministic sequential machine is called P. The set of decision problems that can be solved in polynomial time on a non-deterministic machine is called NP.[2] Equivalently, P is the set of all questions for which there exists an algorithm that can provide an answer in polynomial time and NP is the set of all questions for which an answer to the question can be verified in polynomial time. When talking about NP we rely on this equivalent interpretation as in Cormen et al. (2001), noting that the details are in Hopcroft and Ullman (1979).

Clearly $P \subseteq NP$. Is $P = NP$? No one knows. If it were the case that $P = NP$, then we are essentially saying that a problem that can be quickly checked is also quickly solvable.

Suppose A and B are two decision problems. We say a polynomial-time algorithm f *reduces A to B* if

(i) For any input x of A, $f(x)$ is an input to B; and
(ii) x is a solution to A if and only if $f(x)$ is a solution to B.

If there is a polynomial-time algorithm for B, then there is a polynomial time algorithm for A (Lemma 34.3 in Cormen et al. (2001)). A decision problem A is called *NP-complete* if

(i) A is in NP; and
(ii) Every problem in NP reduces to A via a polynomial-time algorithm.

If the decision problem satisfies (ii) but not necessarily (i), it is called an *NP-hard problem*.

Observe that NP-complete is a subset of NP. Moreover, it is a non-empty subset since one particular decision problem known as the Circuit-Satisfiability Problem has been shown to be an NP-complete problem. The standard technique for showing that a specific decision problem in NP is also NP-complete is to reduce the Circuit-Satisfiability Problem to the specific decision problem via a polynomial-time algorithm (Lemma 34.8 in Cormen et al. (2001)). Many problems have been shown to be NP-complete including the Hamiltonian Cycle Problem.

---

1 The origins of deterministic and non-deterministic machine date back to Alan Turing's paper "On computable numbers, with an application to the Entscheidungsproblem," (Turing, 1937). A good reference book for algorithms is Cormen, Leiserson, Rivest, and Stein's book *Introduction to Algorithms* (Cormen et al., 2001).

2 These concepts appeared in Stephen Cook's 1971 paper "The complexity of theorem proving procedures" (Cook, 1971). However, it is noted in Fortnow (2013) that they appear earlier in a letter by Kurt Gödel to John von Neumann.

If a polynomial time algorithm is found for any one of them, then NP becomes equal to P (Theorem 34.4 in Cormen et al. (2001)).

In 1975 Richard Ladner showed that if P $\neq$ NP, then there exists problems in NP that are neither P nor NP-complete (Ladner, 1975). Such problems are called NP-intermediate. No one has actually found an NP-intermediate problem. It had been conjectured that Graph Isomorphism is *NP-intermediate*. However, recently László Babai posted a preprint on the arXiv[3], where he evidently gave a quasi-polynomial time algorithm for Graph Isomorphism, putting Graph Isomorphism just barely outside P. A quasi-polynomial function has the form $f(n) = 2^{poly(\log n)}$, where *poly*(log n) is the composition of a polynomial function and a log function (a polylogarithmic function). This paper is currently going through the peer-review process.

---

3 "Graph Isomorphism in Quasipolynomial Time", https://arxiv.org/pdf/1512.03547.pdf

Appendix D

Stacks and Queues

A data structure is a way of organizing and storing data in a computer. The graph traversal algorithms described in Section 5.1, Depth-First Search (DFS) and Breadth-First Search (BFS), use two different data structures known as stacks and queues which are described in detail here. This material is taken from Chapters 10 and 13 of Cormen et al. (2001).

A *dynamic set* is a set D that can be manipulated by an algorithm. It can grow, shrink, or change in time. An algorithm can perform two types of operations on dynamic sets: modifying operations that change the set and queries that return information about the set. For example,

- INSERT(D, x) is a modifying operation that augments D by the element x;
- DELETE(D, x) is a modifying operation that removes an element x from D; and
- SEARCH(D, x) is a query that returns TRUE if the element x is in D and FALSE otherwise.

Stacks and queues are dynamic sets in which elements are removed from the set in two different ways. In a *stack* the element deleted is the one most recently inserted. This is like a stack of plates on a spring in a cafeteria: last-in-first-out (LIFO). In a *queue* the element deleted is the one that has been in the set for the longest time. This is like a queue of people: first-in-first-out (FIFO).

A stack with capacity n can be implemented as an array $S[1, \ldots, n]$, together with an attribute $top[S]$, which represents the location of the most recently inserted element or zero when the stack is empty. A stack has two operations, PUSH(S, x) and POP(S). The element $S[1]$ is at the bottom of the stack and the element $S[top[S]]$ is at the top. Executing PUSH(S, x) causes the value x to be stored at location $top[S]$, after which the value of $top[S]$ is incremented by one. If $top[S] = n$, attempting PUSH(S, x) causes a "stack overflow error." Executing POP(S) returns the value stored at position $top[S]$ and decrements $top[S]$ by one. If $top[S] = 0$, executing POP(S) causes a "stack underflow error."

Graphs and Networks, First Edition. S. R. Kingan.

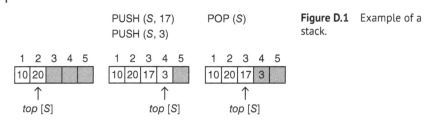

PUSH (*S*, 17)   POP (*S*)   **Figure D.1**   Example of a
PUSH (*S*, 3)   stack.

In the first array in Figure D.1 the stack has capacity 5 and two elements 10 and 20 in positions 1 and 2, respectively. At this stage $top[S] = 2$. The second array shows the stack after operations PUSH(S, 17) and PUSH(S, 3), so at this stage $top[S] = 4$. The third array shows the stack after operation POP(S), and at this stage $top[S] = 3$.

A queue with capacity n can be implemented as an array $Q[1, \dots, n]$ together with two attributes, $head[S]$, which is the index of the next item to remove from the queue reduced modulo n to an index between 1 and n; and $tail[S]$, which is the index of the next item to be stored in the queue, also reduced modulo n. The length of the queue is then $tail[S] - head[S]$.

A queue has two operations: ENQUEUE(Q, x) and DEQUEUE(Q). When ENQUEUE(Q, x) is executed the value x is stored in position $tail[Q]$ (reduced modulo n) and $tail[Q]$ is incremented by 1. When DEQUEUE(Q) is executed, the value at $head[Q]$ (reduced modulo n) is returned and $head[Q]$ is incremented by 1. If ENQUEUE(Q, x) is executed when the length of the queue is n, it causes a "queue overflow error." If DEQUEUE(Q) is executed when the length of the queue is 0, it causes a "queue underflow error."

In the first array in Figure D.2 the queue has capacity 5 and two elements 10 and 20 in positions 3 and 4, respectively. At this stage $head[Q] = 3$ and $tail[Q] = 5$. The second array shows the queue after operations ENQUEUE(Q, 35) and ENQUEUE(Q, 5), so $head[Q] = 3$ and $tail[Q] = 7$. The third array shows the queue after operation DEQUEUE(Q) which returns 10. At this stage $head[Q] = 4$ and $tail[Q] = 7$.

ENQUEUE (*Q*, 35)   DEQUEUE (*Q*)   **Figure D.2**   Example of a
ENQUEUE (*Q*, 5)   queue.

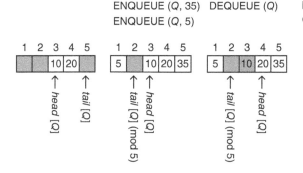

As mentioned in Section 5.1, DFS is a recursive algorithm, whereas BFS is not recursive. Moreover, when DFS encounters a vertex that has not yet been visited, it visits it immediately. To implement a recursive algorithm, a stack is required to record the "state" of the algorithm's data at the point where it executes itself. As BFS examines vertices, it maintains a "to-do" list of vertices that remain to be visited, and visits them in the same order in which they were discovered. That is, it uses a queue. At a deeper level, these two algorithms can be seen as expressions of two related but slightly different ways of organizing a list of tasks. We can use a stack explicitly to rewrite the DFS algorithm in a way that avoids recursion, and demonstrates the relationship between DFS and BFS.

Algorithm D.1. Non-Recursive Depth-First Search

Input: A graph G and a starting vertex s.
Output: An indicator that is TRUE if G is connected and FALSE otherwise and a list T of edges.

(1) Let S be a stack with capacity $|V(G)|$.
(2) $T \leftarrow [\ ]$
(3) $P \leftarrow \{s\}$
(4) $PUSH(S, s)$
(5) **While** $|S| > 0$:
$\qquad v \leftarrow POP(S)$
$\qquad$ **For** each neighbor w of v that is not in P:
$\qquad\qquad P \leftarrow P \cup \{w\}$
$\qquad\qquad T \leftarrow T \cup \{vw\}$
$\qquad\qquad PUSH(S, w)$
(6) **If** $P = V$
$\qquad$ **Return** TRUE
(7) **else**
$\qquad$ **Return** FALSE.

The BFS algorithm is presented again using the precise definition of queue. Note that when DFS is expressed using a stack explicitly, it is almost identical to BFS, with the only significant difference being the use of the stack vs. the queue.

Algorithm D.2. Breadth-First Search (emphasizing queues)

Input: A graph G and a starting vertex s.
Output: An indicator that is TRUE if G is connected and FALSE otherwise and a list T of edges.

(1) Let Q be a queue of size $|V|$

(2) $T \leftarrow [\ ]$

(3) $P \leftarrow \{s\}$

(4) $ENQUEUE(Q, s)$

(5) **While** $|Q| > 0$:

   $v \leftarrow DEQUEUE(Q)$

   For each neighbor w of v that is not in P:

   $P \leftarrow P \cup \{w\}$

   $T \leftarrow T \cup \{vw\}$

   $ENQUEUE(Q, w)$

(6) **If** $P = V$

   Return TRUE

(7) **else**

   Return FALSE.

Bibliography

Adam, D. (2020). A guide to R – the pandemic's misunderstood metric. *Nature*, 583(7816):346–348.

Adamic, L. A. and Adar, E. (2003). Friends and neighbors on the web. *Social Networks*, 25(3):211–230.

Aderem, A. (2005). Systems biology: Its practice and challenges. *Cell*, 121(4):511–513.

Aigner, M. and Ziegler, G. M. (2010). *Proofs from the Book*. Springer, Berlin.

Ajtai, M., Chvátal,V., Newborn, M. M., and Szemerédi, E. (1982). Crossing-free subgraphs. *North-Holland Mathematics Studies*, 60:9–12.

Albert, R., Jeong, H., and Barabási, A.-L. (2000). Error and attack tolerance of complex networks. *Nature*, 406(6794):378–382.

Alekseev, V. B. and Gončakov, V. (1976). The thickness of an arbitrary complete graph. *Mathematics of the USSR – Sbornik*, 30(2):187–202.

Allan, R. B. and Laskar, R. (1978). On domination and independent domination numbers of a graph. *Discrete Mathematics*, 23(2):73–76.

Anderson, I. (1971). Perfect matchings of a graph. *Journal of Combinatorial Theory, Series B*, 10(3):183–186.

Anstee, R. (1982). Properties of a class of (0, 1)-matrices covering a given matrix. *Canadian Journal of Mathematics*, 34(2):438–453.

Archdeacon, D., Ellis-Monaghan, J., Fisher, D., Froncek, D., Lam, P. C. B., Seager, S., Wei, B., and Yuster, R. (2004). Some remarks on domination. *Journal of Graph Theory*, 46(3):207–210.

Backstrom, L., Dwork, C., and Kleinberg, J. (2007). Wherefore art thou R3579X? Anonymized social networks, hidden patterns, and structural steganography. In *Proceedings of the 16th International Conference on World Wide Web*, pages 181–190.

Baker, W. E. and Faulkner, R. R. (1993). The social organization of conspiracy: Illegal networks in the heavy electrical equipment industry. *American Sociological Review*, 58(6):837–860.

Graphs and Networks, First Edition. S. R. Kingan.

Ball, W. W. R. (1892). *Mathematical Recreations and Problems of Past and Present Times*. Macmillan and Company.

Barabási, A.-L. and Albert, R. (1999).Emergence of scaling in random networks. *Science*, 286(5439):509–512.

Barnette, D. (1969). Conjecture 5. In Tutte, W., editor, *Recent Progress in Combinatorics*, New York. Academic Press.

Battle, J., Harary, F., and Kodama, Y. (1962). Every planar graph with nine points has a nonplanar complement. *Bulletin of the American Mathematical Society*, 68(6):569–571.

Bavelas, A. (1950). Communication patterns in task-oriented groups. *Journal of the Acoustical Society of America*, 22(6):725–730.

Bearman, P. S. and Moody, J. (2004). Suicide and friendships among American adolescents. *American Journal of Public Health*, 94(1):89–95.

Beineke, L. and Wilson, R. (2010). The early history of the brick factory problem. *The Mathematical Intelligencer*, 32(2):41–48.

Beineke, L. W. (1970). Characterizations of derived graphs. *Journal of Combinatorial Theory*, 9(2):129–135.

Beineke, L. W. and Chartrand, G. (1968). The coarseness of a graph. *Compositio Mathematica*, 19(4):290–298.

Beineke, L. W., Harary, F., and Moon, J. W. (1964). On the thickness of the complete bipartite graph. In Murphy, T., editor, *Mathematical Proceedings of the Cambridge Philosophical Society*, volume 60, pages 1–5. Cambridge University Press.

Belcastro, S.-M. (2012). The continuing saga of snarks. *The College Mathematics Journal*, 43(1):82–87.

Bellman, R. (1958). On a routing problem. *Quarterly of Applied Mathematics*, 16(1):87–90.

Berge, C. (1957). Two theorems in graph theory. *Proceedings of the National Academy of Sciences of the United States of America*, 43(9):842.

Berge, C. (1958). *La Theorie des Graphes*. Wiley, Paris.

Berge, C. (1963). Perfect graphs. In *Six Papers on Graph Theory*, pages 1–21. Indian Statistical Institute, Calcutta.

Berge, C. (1997). Motivations and history of some of my conjectures. *Discrete Mathematics*, 165:61–70.

Berger, A. (2014). A note on the characterization of digraphic sequences. *Discrete Mathematics*, 314:38–41.

Bielak, H. and Syslo, M. M. (1983). Peripheral vertices in graphs. *Studia Scientiarum Mathematicarum Hungarica*, 18:269–275.

Biggs, N., Lloyd, E. K., and Wilson, R. J. (1976). *Graph Theory, 1736–1936*. Oxford University Press.

Bollobás, B. (2004). *Extremal Graph Theory*. Dover Publications, Inc., Mineola, NY.

Bonacich, P. (1987). Power and centrality: A family of measures. *American Journal of Sociology*, 92(5):1170–1182.

Bondy, J. (1990). Small cycle double covers of graphs. In Hahn, G., Sabidussi, G., and Woodrow, R., editors, *Cycles and Rays*, pages 21–40. Springer, Netherlands.

Bondy, J. A. and Chvátal, V. (1976). A method in graph theory. *Discrete Mathematics*, 15(2):111–135.

Borchardt, C. W. (1861). Über eine interpolationsformel für eine art symmetrischer functionen und über deren anwendung. *Mathematische Abhandlungen der Königlichen Akademie der Wissenschaften zu* Berlin, pages 1–20.

Borüvka, O. (1926a). Contribution to the solution of a problem of economical construction of electrical networks. *Elektronický Obzor*, 15:153–154.

Borüvka, O. (1926b). O jistém problému minimálním (about a certain minimal problem). *Práce Moravské přírodovědecké společnosti*, 3:36–58.

Brin, S. and Page, L. (1998). The anatomy of a large-scale hypertextual web search engine. *Computer Networks and ISDN Systems*, 30(1):107–117.

Broder, A., Kumar, R., Maghoul, F., Raghavan, P., Rajagopalan, S., Stata, R., Tomkins, A., and Wiener, J. (2000). Graph structure in the web. *Computer Networks*, 33(1):309–320.

Brooks, R. L. (1941). On colouring the nodes of a network. *Mathematical Proceedings of the Cambridge Philosophical Society*, 37:194–197.

Buckley, F. and Harary, F. (1990). *Distance in Graphs*. Addison-Wesley, Boston.

Buckley, F., Miller, Z., and Slater, P. J. (1981). On graphs containing a given graph as center. *Journal of Graph Theory*, 5(4):427–434.

Caccetta, L. and Häggkvist, R. (1978). On minimal digraphs with given girth. In *Proceedings of the Ninth Southeastern Conference on Combinatorics, Graph Theory and Computing*, volume XXI, pages 181–187.

Cauchy, A. (1813). Recherche sur les polyèdres-premier mémoire. *Journal de l'Ecole Polytechnique*, 9:66–86.

Cayley, A. (1857). On the theory of the analytical forms called trees. *Philosophical Magazine*, 13(85):172–176.

Cayley, A. (1889). A theorem on trees. *Quarterly Journal of Mathematics*, 23:376–378.

Chaiken, S. and Kleitman, D. J. (1978). Matrix tree theorems. *Journal of Combinatorial Theory, Series A*, 24(3):377–381.

Chartrand, G. (1966). A graph-theoretic approach to a communications problem. *SIAM Journal on Applied Mathematics*, 14(4):778–781.

Chartrand, G. and Harary, F. (1968). Graphs with prescribed connectivities. In Erdős, P. and Katona, G., editors, *Theory of Graphs*, pages 61–63. Academic Press.

Chartrand, G., Kaugars, A., and Lick, D. R. (1972). Critically-connected graphs. *Proceedings of the American Mathematical Society*, 32(1):63–68.

Chartrand, G., Lesniak, L., and Zhang, P. (2011). *Graphs & Digraphs*. CRC Press, Boca Raton, FL, fifth edition.

Chartrand, G. and Tian, S. (1997). Distance in digraphs. *Computers & Mathematics with Applications*, 34(11):15–23.

Chen, Q., Chang, H., Govindan, R., and Jamin, S. (2002). The origin of power laws in internet topologies revisited. In *Proceedings, Twenty-first Annual Joint Conference of the IEEE Computer and Communications Societies*, volume 2, pages 608–617.

Chen, W.-K. (1966). On the realization of a (p, s)-digraph with prescribed degrees. *Journal of the Franklin Institute*, 281(5):406–422.

Chudnovsky, M., Robertson, N., Seymour, P. D., and Thomas, R. (2006). The strong perfect graph theorem. *Annals of Mathematics*, 164(2):51–229.

Chung, F. R. K., Erdős, P., and Graham, R. L. (1981). Minimal decompositions of graphs into mutually isomorphic subgraphs. *Combinatorica*, 1(1):13–24.

Chung, F. R. K., Erdős, P., Graham, R. L., Ulam, S., and Yao, F. (1979). Minimal decompositions of two graphs into pairwise isomorphic subgraphs. *Proceedings, Tenth Southeastern Conference on Combinatorics, Graph Theory and Computing*, 1:3–18.

Chvátal, V. (1972). On Hamilton's ideals. *Journal of Combinatorial Theory, Series B*, 12(2):163–168.

Chvátal, V. (1974). The minimality of the Mycielski graph. In Bari, R. A. and Harary, F., editors, *Graphs and Combinatorics*, pages 243–246. Springer, Berlin.

Coburn, B. J., Wagner, B. G., and Blower, S. (2009). Modeling influenza epidemics and pandemics: insights into the future of swine flu (H1N1). *BMC Medicine*, 7(30):1–8.

Cook, S. A. (1971). The complexity of theorem-proving procedures. In *Proceedings of the Third Annual ACM Symposium on Theory of Computing*, pages 151–158.

Cormen, T., Rivest, R., Leiserson, C., and Stein, C. (2001). *Introduction to Algorithms*. MIT Press, Cambridge, MA.

Costa, L. F., Rodrigues, F. A., and Cristino, A. S. (2008). Complex networks: The key to systems biology. *Genetics and Molecular Biology*, 31(3):591–601.

Coullard, C. R. and Oxley, J. G. (1992). Extensions of Tutte's wheels-and-whirls theorem. *Journal of Combinatorial Theory, Series B*, 56(1):130–140.

Courant, R., Robbins, H., and Stewart, I. (1996). *What is Mathematics? An Elementary Approach to Ideas and Methods*. Oxford University Press, New York.

Csete, M. and Doyle, J. (2004). Bow ties, metabolism and disease. *Trends in Biotechnology*, 22(9):446–450.

Cunningham, D., Everton, S., Wilson, G., Padilla, C., and Zimmerman, D. (2013). Brokers and key players in the internationalization of the FARC. *Studies in Conflict & Terrorism*, 36(6):477–502.

Davis, G. F., Yoo, M., and Baker, W. E. (2003). The small world of the American corporate elite, 1982-2001. *Strategic Organization*, 1(3):301–326.

DeGiuli, E. (2019). Random language model. *Physical Review Letters*, 122(12):128301.

Deo, N. (1974). *Graph Theory with Applications to Engineering and Computer Science*. Dover Publications, Inc., Mineola, NY.

Diestel, R. (2017). *Graph Theory*. Springer, Berlin.

Dijkstra, E. W. (1959). A note on two problems in connexion with graphs. *Numerische Mathematik*, 1(1):269–271.

Dirac, G. A. (1952). Some theorems on abstract graphs. *Proceedings of the London Mathematical Society*, 3(1):69–81.

Dirac, G. A. (1960). 4-chrome graphen und vollständige 4-graphen. *Mathematische Nachrichten*, 22(1-2):51–60.

Doyle, J. and Graver, J. (1977). Mean distance in a graph. *Discrete Mathematics*, 17(2):147–154.

Easley, D. and Kleinberg, J. (2010). *Networks, Crowds, and Markets: Reasoning about a Highly Connected World*. Cambridge University Press.

Edmonds, J. (1965a). Maximum matching and a polyhedron with 0, l-vertices. *Journal of Research of the National Bureau of Standards*, 69(1965):125–130.

Edmonds, J. (1965b). Paths, trees, and flowers. *Canadian Journal of Mathematics*, 17(3):449–467.

Edmonds, J. (1991). A glimpse of heaven. In Lenstra, J., Kan, A. R., and Schrijver, A., editors, *History of Mathematical Programming — A Collection of Personal Reminiscences*, pages 32–54. North-Holland, Amsterdam.

Edmonds, J. and Johnson, E. L. (1973). Matching, Euler tours and the Chinese postman. *Mathematical Programming*, 5(1):88–124.

Edwards, K., Sanders, D. P., Seymour, P. D., and Thomas, R. (2016). Three-edge-colouring doublecross cubic graphs. *Journal of Combinatorial Theory, Series B*, 119:66–95.

Egerváry, E. (1931). On combinatorial properties of matrices. *Matematikai és Fizikai Lapok*, 38:16–28.

Ellingham, M. N. and Horton, J. D. (1983). Non-Hamiltonian 3-connected cubic bipartite graphs. *Journal of Combinatorial Theory, Series B*, 34(3):350–353.

Erdős, P. and Rényi, A. (1959). On random graphs I. *Publicationes Mathematicae Debrecen*, 6:290–297.

Erdős, P. and Gallai, T. (1960). Gráfok előírt fokszámú pontokkal. *Matematikai és Fizikai Lapok*, 11:264–274.

Erdős, P., Pach, J., Pollack, R., and Tuza, Z. (1989). Radius, diameter, and minimum degree. *Journal of Combinatorial Theory, Series B*, 47(1):73–79.

Euler, L. (1736). Solutio promblematis ad geometriam sirus pertinentis. *Commentarii Academiae Scientiarum Imperialis Petropolitanae*, 8:128–140.

Euler, L. (1752). Elementa doctrinae solidorum, novi commentarii academiae scientiarum imperialis petropolitanae. *Novi Commentarii Academiae Scientiarum Imperialis Petropolitanae*, 4:109–140.

Everton, S. (2012). *Disrupting Dark Networks*. Cambridge University Press.

Fallat, S. M., Hogben, L., Lin, J. C.-H., and Shader, B. L. (2020). The inverse eigenvalue problem of a graph, zero forcing, and related parameters. *Notices of the American Mathematical Society*, 67(2):257–261.

Faulkner, G. B. and Younger, D. (1974). Non-Hamiltonian cubic planar maps. *Discrete Mathematics*, 7(1-2):67–74.

Faulon, J., Visco, D. P., and Roe, D. (2005). Enumerating molecules. *Reviews in Computational Chemistry*, 21:209–286.

Fiedler, M. (1973). Algebraic connectivity of graphs. *Czechoslovak Mathematical Journal*, 23(2):298–305.

Floyd, R. W. (1962). Algorithm 97: Shortest path. *Communications of the ACM*, 5(6):345.

Ford, L. R. and Fulkerson, D. R. (1956). Maximal flow through a network. *Canadian Journal of Mathematics*, 8(3):399–404.

Ford, L. R. and Fulkerson, D. R. (1962). *Flows in networks*. Princeton University Press.

Fortnow, L. (2013). *The Golden Ticket: P, NP, and the Search for the Impossible*. Princeton University Press.

Foulds, L., Perera, S., and Robinson, D. (1978). Network layout procedure for printed-circuit design. *Computer-Aided Design*, 10(3):177–180.

Foulds, L. R. (2012). *Graph Theory Applications*. Springer Science & Business Media, New York.

Freeman, L. C. (1979). Centrality in social networks conceptual clarification. *Social Networks*, 1(3):215–239.

Fujie, F. and Zhang, P. (2014). *Covering Walks in Graphs*. Springer, New York.

Fulkerson, D. R. (1960). Zero-one matrices with zero trace. *Pacific Journal of Mathematics*, 10(3):831–836.

Gale, D. and Shapley, L. S. (1962). College admissions and the stability of marriage. *American Mathematical Monthly*, 69(1):9–15.

Gallai, T. (1958). Maximum-minimum sätze über graphen. *Acta Mathematica Hungarica*,9(3-4):395–434.

Gans, H. J. (1962). *The Urban Villagers: Group and Class in the Life of Italians-Americans*. Free Press of Glencoe, New York.

Gans, H. J. (1974). Gans on Granovetter's "Strength of weak ties". *American Journal of Sociology*, 80(2):524–527.

Garcia-Lopez, P., Garcia-Marin, V., and Freire, M. (2010). The histological slides and drawings of Cajal. *Frontiers in Neuroanatomy*, 4:9.

Gardner, M. (1976). Mathematical games. *Scientific American*, 234:126–130.

Garey, M. R. and Johnson, D. S. (1983). Crossing number is NP-complete. *SIAM Journal on Algebraic Discrete Methods*, 4(3):312–316.

Ge, Y., Zhao, S., Zhou, H., Pei, C., Sun, F., Ou, W., and Zhang, Y. (2020). Understanding echo chambers in e-commerce recommender systems. In

Proceedings of the 43rd International ACM SIGIR Conference on Research and Development in Information Retrieval, pages 2261–2270.

Georges, J. P. (1989). Non-Hamiltonian bicubic graphs. *Journal of Combinatorial Theory, Series B*, 46(1):121–124.

Gethner, E., Hogben, L., Lidický, B., Pfender, F., Ruiz, A., and Young, M. (2017). On crossing numbers of complete tripartite and balanced complete multipartite graphs. *Journal of Graph Theory*, 84(4):552–565.

Gilbert, E. N. (1959). Random graphs. *The Annals of Mathematical Statistics*, 30(4):1141–1144.

Giot, L., Bader, J. S., Brouwer, C., Chaudhuri, A., Kuang, B., Li, Y., Hao, Y., Ooi, C., Godwin, B., Vitols, E., et al. (2003). A protein interaction map of Drosophila melanogaster. *Science*, 302(5651):1727–1736.

Goddard, W. and Oellermann, O. R. (2011). Distance in graphs. In Dehmer, M., editor, *Structural Analysis of Complex Networks*, pages 49–72. Springer Science & Business Media, New York.

Gould, R. J. (1991). Updating the Hamiltonian problem–a survey. *Journal of Graph Theory*, 15(2):121–157.

Gould, R. J. (2003). Advances on the Hamiltonian problem–a survey. *Graphs and Combinatorics*, 19(1):7–52.

Graham, R. L. (1987). Similarity measure for graphs. *Los Alamos Science*, 15:114–121.

Graham, R. L. and Pollak, H. O. (1971). On the addressing problem for loop switching. *Bell System Technical Journal*, 50(8):2495–2519.

Graham, R. L. and Pollak, H. O. (1972). On embedding graphs in squashed cubes. In Alavi, Y., Lick, D., and White, A., editors, *Graph Theory and Applications*, pages 99–110. Springer.

Granovetter, M. (1974). *Getting a Job: A Study of Contacts and Careers*. University of Chicago Press.

Granovetter, M. (1983). The strength of weak ties: A network theory revisited. *Sociological Theory*, 1(1):201–233.

Granovetter, M. S. (1973). The strength of weak ties. *American Journal of Sociology*, 78(6):1360–1380.

Grinberg, É. J. (1968). Plane homogeneous graphs of degree three without Hamiltonian circuits. *Latvian Mathematical Yearbook*, 4:51–58.

Gross, J. L. and Yellen, J. (2005). *Graph Theory and its Applications*. CRC Press, Boca Raton, FL.

Grossman, J. W. and Ion, P. D. (1995). On a portion of the well-known collaboration graph. In *Proceedings of the Twenty-sixth Southeastern International Conference on Combinatorics, Graph Theory and Computing*, volume 108, pages 129–132.

Grötschel, M. and Yuan, Y. (2012). Euler, Mei-Ko Kwan, Königsberg, and a Chinese postman. *Documenta Mathematica*, Extra Volume: Optimization Stories:43–50.

Grünbaum, B. (2003). *Convex Polytopes*. Springer Science & Business Media, New York.

Grünbaum, B. (2007). Graphs of polyhedra; polyhedra as graphs. *Discrete Mathematics*, 307(3-5):445–463.

Guy, R. K. (1960). A combinatorial problem. *Nabla (Bulletin of the Malayan Mathematical Society)*, 7:68–72.

Guy, R. K. (1969). The decline and fall of Zarankiewicz's theorem. In Harary, F., editor, *Proof Techniques in Graph Theory*, pages 63–69. Academic Press, New York.

Hadwiger, H. (1943). Über eine klassifikation der streckenkomplexe. *Vierteljahrsschrift der Naturforschenden Gesellschaft in Zürich*, 88:133–142.

Hakimi, S. L. (1962). On realizability of a set of integers as degrees of the vertices of a linear graph. I. *Journal of the Society for Industrial & Applied Mathematics*, 10(3):496–506.

Hall, P. (1935). On representatives of subsets. *Journal of the London Mathematical Society*, 10(1):26–30.

Hamilton, W. R. (1858). Account of the Icosian calculus. In Lloyd, H., editor, *Proceedings of the Royal Irish Academy*, volume 6, pages 415–416. Royal Irish Academy.

Harary, F. (1959). Status and contrastatus. *Sociometry*, 22(1):23–43.

Harary, F. (1961). Research problem. *Bulletin of the American Mathematical Society*, 67:542.

Harary, F. (1964). On the reconstruction of a graph from a collection of subgraphs. In *Theory of Graphs and its Applications: Proceedings of the Symposium held in Smolenice in June 1963*, pages 47–52.

Harary, F. and Norman, R. Z. (1953). The dissimilarity characteristic of Husimi trees. *Annals of Mathematics*, 58(1):134–141.

Harper, F. M. and Konstan, J. A. (2015). The movielens datasets: History and context. *ACM Transactions on Interactive Intelligent Systems*, 5(4):1–19.

Havel, V. (1955). A remark on the existence of finite graphs. *Casopis pro Pestování Matematiky*, 80:477–480.

Haynes, T. W., Hedetniemi, S., and Slater, P. (1998). *Fundamentals of Domination in Graphs*. CRC Press, Boca Raton, FL.

Heawood, P. J. (1890). Map colour theorems. *Quarterly Journal of Mathematics, Series 2*, 24:332–338.

Hendry, G. R. (1985). On graphs with prescribed median I. *Journal of Graph Theory*, 9(4):477–481.

Hierholzer, C. (1873). Über die möglichkeit, einen linienzug ohne. *Mathematische Annalen*, 6:30–32.

Holbert, K. S. (1989). A note on graphs with distant center and median. In Kulli, V., editor, *Recent Sudies in Graph Theory*, pages 155–158, Gulbarza, India.

Hopcroft, J. E. and Ullman, J. D. (1979). *Introduction to Automata Theory, Languages, and Computation*. Addison-Wesley, Boston, second edition.

Hopkins, B. and Wilson, R. J. (2004). The truth about Königsberg. *The College Mathematics Journal*, 35(3):198–207.

Horn, R. A. and Johnson, C. R. (1990). *Matrix Analysis*. Cambridge University Press.

Horton, J. D. (1982). On two-factors of bipartite regular graphs. *Discrete Mathematics*, 41(1):35–41.

István, F. (1948). On straight-line representation of planar graphs. *Acta Scientiarum Mathematicarum*, 11:229–233.

Jaccard, P. (1901). Etude comparative de la distribution florale dans une portion des alpes et du jura. *Bulletin de la Societe Vaudoise des Sciences Naturelles*, 37:547–579.

Jeong, H., Mason, S. P., Barabási, A.-L., and Oltvai, Z. N. (2001). Lethality and centrality in protein networks. *Nature*, 411(6833):41–42.

Jordan, C. (1869). Sur les assemblages de lignes. *Journal für die Reine und Angewandte Mathematik*, 70:185–190.

Joyal, A. (1981). Une théorie combinatoire des séries formelles. *Advances in Mathematics*, 42(1):1–82.

Karp, R. M. (1972). Reducibility among combinatorial problems. In Miller R. E., Thatcher J. W., B. J. D., editor, *Complexity of Computer Computations*. Springer, Boston.

Katz, L. (1953). A new status index derived from sociometric analysis. *Psychometrika*, 18(1):39–43.

Kelly, P. J. (1957). A congruence theorem for trees. *Pacific Journal of Mathematics*, 7(1):961–968.

Kempe, A. B. (1879). On the geographical problem of the four colours. *American Journal of Mathematics*, 2(3):193–200.

Kempe, A. B. (1886). A memoir on the theory of mathematical form. *Philosophical Transactions of the Royal Society of London*, 177:1–70.

Kempe, D., Kleinberg, J., and Kumar,A. (2002). Connectivity and inference problems for temporal networks. *Journal of Computer and System Sciences*, 64(4):820–842.

Kirchhoff, G. (1847). Uberdie auflosung der gleichungen, auf welche man bei der untersuchung der linearen verteilung galvanischer strome geluhrt wird. *Annalen der Physik und Cheme*, 72:497–508.

Kirkman, T. P. (1856). On the representation of polyhedra. *Philosophical Transactions of the Royal Society*, 146:313–418.

Kleinberg, J. M. (1999). Authoritative sources in a hyperlinked environment. *Journal of the ACM*, 46(5):604–632.

König, D. (1931). Gráfok és mátrixok. *Matematikai és Fizikai Lapok*, 38:116–119.

König, D. (1936). *Theorie der endlichen und unendlichen Graphen*. Akademische Verlagsgesellschaft.

Krausz, J. (1943). Démonstration nouvelle d'une théorème de Whitney sur les réseaux. *Matematikai és Fizikai Lapok*, 50:75–85.

Kruskal, J. B. (1956). On the shortest spanning subtree of a graph and the traveling salesman problem. *Proceedings of the American Mathematical Society*, 7(1):48–50.

Kuhn, H. W. (1955). The Hungarian method for the assignment problem. *Naval Research Logistics Quarterly*, 2(1-2):83–97.

Kupferschmidt, K. (2020). Case clustering emerges as key pandemic puzzle. *Science*, 368(6493):808–809.

Kuratowski, K. (1930). Sur le probleme des courbes gauches en topologie. *Fundamenta Mathematicae*, 15(1):271–283.

Kwan, M.-K. (1962). Graphic programming using odd or even points. *Chinese Mathematics*, 1:(273–277).

Ladner, R. E. (1975). On the structure of polynomial time reducibility. *Journal of the ACM*, 22(1):155–171.

Lay, D. C. (2016). *Linear Algebra and its applications*. Addison-Wesley, Boston, fourth edition.

Leicht, E., Holme, P., and Newman, M. E. (2006). Vertex similarity in networks. *Physical Review E*, 73(2):026120.

Leighton, F. T. (1983). *Complexity Issues in VLSI: Optimal Layouts for the Shuffle-Exchange Graph and other Networks*. MIT press.

Li, A., Cornelius, S. P., Liu, Y.-Y., Wang, L., and Barabási, A.-L. (2017). The fundamental advantages of temporal networks. *Science*, 358(6366):1042–1046.

Lovász, L. (1968). On covering of graphs. In Erdős, P. and Katona, G., editors, *Theory of Graphs: Proceedings of the Colloquium Held at Tihany, Hungary, September, 1966*, pages 231–236. Academic Press New York.

Lovász, L. (1972). Normal hypergraphs and the perfect graph conjecture. *Discrete Mathematics*, 2(3):253–267.

Lovász, L. (1975). Three short proofs in graph theory. *Journal of Combinatorial Theory, Series B*, 19(3):269–271.

Lusseau, D., Schneider, K., Boisseau, O. J., Haase, P., Slooten, E., and Dawson, S. M. (2003). The bottlenose dolphin community of Doubtful Sound features a large proportion of long-lasting associations. *Behavioral Ecology and Sociobiology*, 54(4):396–405.

Mäkinen, E., Poranen, T., et al. (2012). An annotated bibliography on the thickness, outerthickness, and arboricity of a graph. *Missouri Journal of Mathematical Sciences*, 24(1):76–87.

Martinez, N. D. (1991). Artifacts or attributes? Effects of resolution on the Little Rock Lake food web. *Ecological Monographs*, 61(4):367–392.

Melnikov, L. and Vizing, V. (1969). New proof of Brooks' theorem. *Journal of Combinatorial Theory*, 7(4):289–290.

Menger, K. (1927). Zur allgemeinen kurventheorie. *Fundamenta Mathematicae*, 10(1):96–115.

Metelsky, Y. and Tyshkevich, R. (1997). On line graphs of linear 3-uniform hypergraphs. *Journal of Graph Theory*, 25(4):243–251.

Meusel, R., Vigna, S., Lehmberg, O., and Bizer, C. (2015). The graph structure in the web: Analyzed on different aggregation levels. *The Journal of Web Science*, 1(1):33–47.

Milgram, S. (1963). Behavioral study of obedience. *The Journal of Abnormal and Social Psychology*, 67(4):371–378.

Miller, I. and Miller, M. (2014). *Freund's Mathematical Statistics with Applications*. Prentice Hall, Saddle River, NJ.

Mohar, B., Alavi, Y., Chartrand, G., and Oellermann, O. (1991). The Laplacian spectrum of graphs. *Graph Theory, Combinatorics, and Applications*, 2:871–898.

Moon, J. (1970). Counting Labelled Trees. Canadian Mathematical Congress, Ottowa, CA.

Moreno, J. L. (1934). *Who shall survive? Foundations of Sociometry, Group Psychotherapy and Sociodrama*. Nervous and Mental Disease Publishing Co., Washington, D.C.

Mulder, H. M. (1992). Julius Petersen's theory of regular graphs. *Discrete Mathematics*, 100(1):157–175.

Mutzel, P., Odenthal, T., and Scharbrodt, M. (1998). The thickness of graphs: A survey. *Graphs and Combinatorics*, 14(1):59–73.

Mycielski, J. (1955). Sur le coloriage des graphes. *Colloquium Mathematicum*, 3:161–162.

Negami, S. (1982). A characterization of 3-connected graphs containing a given graph. *Journal of Combinatorial Theory, Series B*, 32(1):69–74.

Nešetřil, J., Milková, E., and Nešetřilová, H. (2001). Otakar Borůvka on minimum spanning tree problem translation of both the 1926 papers, comments, history. *Discrete Mathematics*,233(1-3):3–36.

Newman, M. E. (2001). The structure of scientific collaboration networks. *Proceedings of the National Academy of Sciences*, 98(2):404–409.

Newman, M. E. (2002). Assortative mixing in networks. *Physical Review Letters*, 89(20):208701.

Newman, M. E. (2003a). Mixing patterns in networks. *Physical Review E*, 67(2):026126.

Newman, M. E. (2003b). The structure and function of complex networks. *SIAM Review*, 45(2):167–256.

Newman, M. E. (2005). Power laws, Pareto distributions and Zipf's law. *Contemporary Physics*, 46(5):323–351.

Novotny, K. and Tian, S. (1991). On graphs with intersecting center and median. In Kulli, V., editor, *Advances in Graph Theory*, pages 297–300. Vishwa International Publications, Gulbarga, India.

Ore, O. (1960). Note on Hamilton circuits. *American Mathematical Monthly*, 67:55–55.

Oxley, J. G. (2011). *Matroid Theory*. Oxford University Press, second edition.

Pach, J. and Sharir, M. (2009). *Combinatorial Geometry and its Algorithmic Applications*: The Alcalá lectures. American Mathematical Society, Providence, RI.

Padgett, J. F. and Ansell, C. K. (1993). Robust action and the rise of the Medici, 1400-1434. *American Journal of Sociology*, 98(6):1259–1319.

Pareto, V. (1964). *Cours D'économie Politique*. Librairie Droz, Geneva.

Parks, D. J. (2012). *Graph Theory in America 1876-1950*. PhD thesis, The Open University.

Pastor-Satorras, R., Vázquez, A., and Vespignani, A. (2001). Dynamical and correlation properties of the internet. *Physical Review Letters*, 87(25):258701.

Petersen, J. (1891). Die theorie der regulären graphs. *Acta Mathematica*, 15(1):193–220.

Petersen, J. (1898). Sur le théoreme de tait. *L'intermédiaire des Mathématiciens*, 5:225–227.

Powers, D. M. (1998). Applications and explanations of Zipf's law. In *Proceedings of the Joint Conferences on New Methods in Language Processing and Computational Natural Language Learning*, pages 151–160. Association for Computational Linguistics.

Price, D. J. D. S. (1965). Networks of scientific papers. *Science*, 149(3683):510–515.

Prim, R. C. (1957). Shortest connection networks and some generalizations. *Bell System Technical Journal*, 36(6):1389–1401.

Pulleyblank, W. R. (2012). Edmonds, matching and the birth of polyhedral combinatorics. *Documenta Mathematica*, pages 181–197.

Rain, J.-C., Selig, L., De Reuse, H., Battaglia, V., Reverdy, C., Simon, S., Lenzen, G., Petel, F., Wojcik, J., Schächter, V., et al. (2001). The protein–protein interaction map of helicobacter pylori. *Nature*, 409(6817):211–215.

Ramsey, F. P. (1930). On a problem of formal logic. *Proceedings of the London Mathematical Society*, s2-30 (1):264–286.

Reed, B. (1996). Paths, stars and the number three. *Combinatorics, Probability & Computing*, 5:277–295.

Rényi, A. (1970). On the enumeration of trees. In Guy, R., editor, *Combinatorial Structures and their Applications*, pages 355–360, London. Gorden and Breach.

Ringel, G. (1965). Das geschlecht des vollständigen paaren graphen. In Bauer, H., Collatz, L., Hasse, H., Kähler, E., Sperner, E., and Witt, E., editors, *Abhandlungen aus dem Mathematischen Seminar der Universität Hamburg*, pages 139–150. Springer, Berlin.

Robbins, H. (1939). A theorem on graphs, with an application to a problem of traffic control. *The American Mathematical Monthly*, 46(5):281–283.

Roberts, F. (2005). *Applied Combinatorics*. Prentice-Hall, Upper Saddle River, NJ.

Roberts, N. and Everton, S. F. (2011). Strategies for combating dark networks. *Journal of Social Structure*, 12:1–32.

Robertson, N., Sanders, D., Seymour, P. D., and Thomas, R. (1997a). The four-colour theorem. *Journal of Combinatorial Theory, Series B*, 70(1):2–44.

Robertson, N. and Seymour, P. D. (1983). Graph minors I. Excluding a forest. *Journal of Combinatorial Theory, Series B*, 35(1):39–61.

Robertson, N. and Seymour, P. D. (1986). Graph minors V. Excluding a planar graph. *Journal of Combinatorial Theory, Series B*, 41(1):92–114.

Robertson, N. and Seymour, P. D. (2003). Graph minors XVI. Excluding a non-planar graph. *Journal of Combinatorial Theory, Series B*, 89(1):43–76.

Robertson, N. and Seymour, P. D. (2004). Graph minors XX. Wagner's conjecture. *Journal of Combinatorial Theory, Series B*, 92(2):325–357.

Robertson, N. and Seymour, P. D. (2010). Graph minors XXIII. Nash-Williams' immersion conjecture. *Journal of Combinatorial Theory, Series B*, 100(2):181–205.

Robertson, N., Seymour, P. D., and Thomas, R. (1993). Hadwiger's conjecture for K_6-free graphs. *Combinatorica*, 13(3):279–361.

Robertson, N., Seymour, P. D., and Thomas, R. (1997b). Tutte's edge-colouring conjecture. *Journal of Combinatorial Theory, Series B*, 70(1):166–183.

Saaty, T. L. (1964). The minimum number of intersections in complete graphs. *Proceedings of the National Academy of Sciences*, 52(3):688–690.

Schlegel, V. (1886). Über projectionsmodelle der regelmässigen vier-dimensionalen körper.

Schoch, D. and Brandes, U. (2016). Re-conceptualizing centrality in social networks. *European Journal of Applied Mathematics*, 27(6):971–985.

Schrijver, A. (2002). On the history of the transportation and maximum flow problems. *Mathematical Programming*, 91(3):437–445.

Schrijver, A. (2003). *Combinatorial Optimization, volume A*. Springer, Berlin.

Seeley, J. R. (1949). The net of reciprocal influence; a problem in treating sociometric data. *Canadian Journal of Psychology*, 3:234–240.

Seung, S. (2012). *Connectome: How the Brain's Wiring Makes Us Who We Are*. Houghton Mifflin Harcourt, Boston.

Seymour, P. D. (1980). Decomposition of regular matroids. *Journal of Combinatorial Theory Series B*, 28:305–359.

Shapiro, A. (2017). Reform predictive policing. *Nature News*, 541(7638):458.

Shitov, Y. (2019). Counterexamples to Hedetniemi's conjecture. *Annals of Mathematics*, 190(2):663–667.

Sipka, T. (2002). Alfred Bray Kempe's "proof" of the Four-Color Theorem. *Math Horizons*, 10(2):21–26.

Slater, P. J. (1980). Medians of arbitrary graphs. *Journal of Graph Theory*, 4(4):389–392.

Sparrow, M. K. (1991). The application of network analysis to criminal intelligence: An assessment of the prospects. *Social Networks*, 13(3):251–274.

Steinitz, E. (1922). Polyeder und raumeinteilungen. *Encyclopädie der Mathematischen Wissenschaften*, 12:38–43.

Stockmeyer, P. K. (1977). The falsity of the reconstruction conjecture for tournaments. *Journal of Graph Theory*, 1(1):19–25.

Sylvester, J. J. (1878a). Chemistry and algebra. *Nature*, 17:284.

Sylvester, J. J. (1878b). On an application of the new atomic theory to the graphical representation of the invariants and covariants of binary quantics, with three appendices. *American Journal of Mathematics*, 1(1):64–104.

Szekeres, G. (1973). Polyhedral decompositions of cubic graphs. *Bulletin of the Australian Mathematical Society*, 8(03):367–387.

Szekeres, G. and Wilf, H. S. (1968). An inequality for the chromatic number of a graph. *Journal of Combinatorial Theory*, 4(1):1–3.

Tait, P. (1884). IV. Listings topologie. *Philosophical Magazine*, 17(103):30–46.

Thorpe, T. (1910). *History of Chemistry: From 1850 to 1910*. Watts & Company, London.

Tripathi, A., Venugopalan, S., and West, D. B. (2010). A short constructive proof of the Erdős–Gallai characterization of graphic lists. *Discrete Mathematics*, 310(4):843–844.

Tsonis, A. A., Swanson, K. L., and Roebber, P. J. (2006). What do networks have to do with climate? *Bulletin of the American Meteorological Society*, 87(5):585–596.

Turán, P. (1977). A note of welcome. *Journal of Graph Theory*, 1(1):7–9.

Turing, A. M. (1937). On computable numbers, with an application to the entscheidungsproblem. *Proceedings of the London Mathematical Society*, 2(1):230–265.

Tutte, W. T. (1946). On Hamiltonian circuits. *Journal of the London Mathematical Society*, 1(2):98–101.

Tutte, W. T. (1947). The factorization of linear graphs. *Journal of the London Mathematical Society*, 1(2):107–111.

Tutte, W. T. (1956). A theorem on planar graphs. *Transactions of the American Mathematical Society*, 82:99–116.

Tutte, W. T. (1961). A theory of 3-connected graphs. *Indagationes Mathematicae*, 23:441–455.

Tutte, W. T. (1963a). The non-biplanar character of the complete 9-graph. *Canadian Mathematical Bulletin*, 6(3):319–330.

Tutte, W. T. (1963b). The thickness of a graph. *Indagationes Mathematicae*, 25:567–577.

Tutte, W. T. (1971). On the 2-factors of bicubic graphs. *Discrete Mathematics*, 1(2):203–208.

Ulam, S. M. (1960). *A Collection of Mathematical Problems*. Interscience, New York.

Veblen, O. (1912). An application of modular equations in analysis situs. *Annals of Mathematics*, 14(1/4):86–94.

Veblen, O. and Evans, G. C. (1922). *The Cambridge Colloquium 1916. Part II: Analysis Situs*. American Mathematical Society, Providence, RI.

Vizing, V. (1964). On an estimate of the chromatic class of a p-graph. *Diskretnyi Analiz i Issledovanie Operatsii*, 3:25–30.

von Staudt, K. G. C. (1847). *Geometrie der Lage*. Bauer u. Raspe, Nürnberg.

Wagner, K. (1937). Über eine Erweiterung eines Satzes von Kuratowski. *Deutsche Mathematik*, 2:280–285.

Warshall, S. (1962). A theorem on boolean matrices. *Journal of the ACM*, 9(1):11–12.

Watts, D. J. and Strogatz, S. H. (1998). Collective dynamics of 'small-world' networks. *Nature*, 393(6684):440–442.

West, D. B. (2001). *Introduction to Graph Theory*, Prentice Hall, Upper Saddle River, NJ, second edition.

Whitney, H. (1932a). Congruent graphs and the connectivity of graphs. *American Journal of Mathematics*, 54(1):150–168.

Whitney, H. (1932b). Non-separable and planar graphs. *Transactions of the American Mathematical Society*, 34 (2):339–362.

Whitney, H. (1933). 2-isomorphic graphs. *American Journal of Mathematics*, 55(1):245–254.

Whitney, H. (1935). On the abstract properties of linear dependence. *American Journal of Mathematics*, 57(3):509–533.

Wiedermann, M., Radebach, A., Donges, J. F., Kurths, J., and Donner, R. V. (2016). A climate network-based index to discriminate different types of El Niño and La Niña. *Geophysical Research Letters*, 43(13):7176–7185.

Wiener, H. (1947). Structural determination of paraffin boiling points. *Journal of the American Chemical Society*, 69(1):17–20.

Willinger, W., Alderson, D., and Doyle, J. C. (2009). *Mathematics and the Internet: A source of enormous confusion and great potential*. Defense Technical Information Center.

Winkler, P. M. (1983). Proof of the squashed cube conjecture. *Combinatorica*, 3(1):135–139.

Witzgall, C. (2001). Paths, trees, and flowers. In Lide, D. R., editor, *A Century of Excellence in Measurements, Standards, and Technology*, pages 140–144. National Institute of Standards and Technology.

Zachary, W. W. (1977). An information flow model for conflict and fission in small groups. *Journal of Anthropological Research*, 33(4):452–473.

Zarankiewicz, C. (1955). On a problem of P. Turán concerning graphs. *Fundamenta Mathematicae*, 41(1):137–145.

Zipf, G. K. (1935). *The Psycho-Biology of Language; an Introduction to Dynamic Philology*. Routledge, London.

Index

Note: Page numbers in bold denote theorems and algorithms

Graphs and Networks, First Edition. S. R. Kingan.
© 2022 John Wiley & Sons, Inc. Published 2022 by John Wiley & Sons, Inc.